Groundborne Noise

This book is for practitioners in the measurement, prediction, assessment and mitigation of groundborne noise, as well as acoustics engineering students. It explains the sources of groundborne noise, how they transmit into the medium supporting them (or in which they are embedded), the range of wave types that result, how they are propagated through all the different forms of soil, rock or building structures that connect them to receptors, and how the vibration that is groundborne noise is transmitted into receiving structures, to be re-radiated as an acoustical phenomenon in rooms and other spaces. It is clear, readable and logical, with enough technical detail for an engineer without being needlessly complex.

Groundborne Noise

Rupert Thornely-Taylor

CRC Press
Taylor & Francis Group
Boca Raton London New York

CRC Press is an imprint of the
Taylor & Francis Group, an **informa** business

Designed cover image: Image courtesy of the author

First edition published 2027
by CRC Press
2385 NW Executive Center Drive, Suite 320, Boca Raton FL 33431

and by CRC Press
4 Park Square, Milton Park, Abingdon, Oxon, OX14 4RN

CRC Press is an imprint of Taylor & Francis Group, LLC

ISBN: 978-1-041-01997-8 (hbk)
ISBN: 978-1-041-02328-9 (pbk)
ISBN: 978-1-003-61877-5 (ebk)

DOI: 10.1201/9781003618775

Typeset in Sabon
by SPi Technologies India Pvt Ltd (Straive)

Contents

Preface

There is a large body of literature on the subject of airborne acoustics, but a phenomenon that results in airborne sound in limited but important circumstances is not well documented, namely groundborne noise. This is sound, or noise, that is perceived by the sense of hearing, but which reaches the air in which the audible stimulus occurs by transmission through the ground, or through other solid media including structures, from a source of vibration at acoustic frequencies.

Because solid media are not just elastic, but also possess shear stiffness, many more forms of wave propagation occur than is the case with air, and the entire process, from characterising the source through to predicting the waves it emits and the manner in which they are propagated and received in the location where they are eventually heard as sound, is of a complexity much greater than in the case of purely airborne noise.

The acoustics profession now has many members, and there are also many specialists who work in vibration, but when they are faced with problems of groundborne or structureborne noise they find themselves in somewhat unfamiliar territory. They are thrown back on to a small number of sources of empirical or simple guidance, without necessarily understanding the magnitude of the uncertainty associated with it. In some cases the responsibility when giving professional advice on the likely levels of groundborne noise, the acceptability of those levels, and ways of mitigating unacceptable levels can be very great. It is common now for developers to seek to develop sites which are, for example, close to or above railways, and the mitigation necessary is to mount the entire building on resilient bearings. This brings not only cost but also risk and a need for particular structural engineering skills. Dealing with problems of groundborne noise in new buildings retrospectively, if an approach has been taken that was not well-enough informed, can be almost impossible.

This book seeks to take acoustical engineers onward from their original familiar territory into the elaborate world of waves from unfamiliar sources travelling through unfamiliar media. Because of the complexities of wave motion in solid media, there are many cases where text-book solutions are not possible because the equations are too intractable and the only realistic way forward is by means of numerical modelling and computer simulation. While that takes engineers onto a further stage, which can only be outlined in a book, it is, nevertheless, most unwise to embark on numerical approaches to dynamic problems without the benefit of a full understanding of what is going on in computer models, not least in order to be able to foresee where potential mitigation measures may lead.

A note on the author

Rupert Thornely-Taylor is a Fellow of, and was a founder member of, the Institute of Acoustics, a Member of the Institute of Noise Control Engineering of the USA and a Fellow of the International Institute of Acoustics and Vibration. He has specialised exclusively in the subjects of noise, vibration and acoustics for more than 60 years. He has been an independent consultant in these subjects for the past 58 years and heads the Rupert Taylor Ltd consultancy practice. He is a past President and Honorary Member of the Association of Noise Consultants and a past Director of the International Institute of Acoustics and Vibration. He is the author of the Pelican books *Noise and Electricity*, and editor or co-author of many other books including the Association of Noise Consultants Guidelines *Measurement & Assessment of Groundborne Noise & Vibration* He is a member of the working group that produced ISO Standard 14837-1:2005 "Mechanical vibration - Ground-borne noise and vibration arising from rail systems - Part 1: General guidance" and ISO Standard 14837-32:2015 "Measurement of dynamic properties of the ground". He has prepared reports on noise for the OECD. In 2016 the Institute of Acoustics awarded him its Rayleigh Medal for outstanding contributions to acoustics. In 2013 the Association of Noise Consultants awarded him their Outstanding Contribution Award. A high proportion of his work has been devoted to groundborne noise and vibration, which was the subject of his Rayleigh Medal lecture, and his plenary presentation at the 26th International Congress on Sound and Vibration in Montreal in 2019. He developed the FINDWAVE Finite-Difference-Time-Domain computer package which has been used for simulations of groundborne noise and vibration of many major developments including High Speed 2 and a wide range of buildings near railways and other sources of vibration in the UK, Spain, Scandinavia, Singapore and China.

Chapter 1

Introduction

Groundborne noise is sound that that reaches a receptor after travelling via a path that, for all but the last part, contains a medium other than air. It is audible by the human ear, as is any sound, and once in the air around where it is heard, it is subject to all the usual physics of airborne acoustics. There are cases where groundborne noise does not pass through air at all, for example sensitive equipment may be affected if vibration reaches it through its support and foundation.

This book concentrates on groundborne noise rather than vibration, even though the former is a manifestation of the latter. There is inevitably some overlap between the two topics, but consideration is not specifically given to vibration in terms of a phenomenon that is primarily perceived by the sense of touch or motion, or which can cause physical effects on structures such as damage.

The transmission path of groundborne noise is not the only feature that distinguishes it from normal airborne sound. In most environments, no groundborne noise can be heard, whereas only in special cases is no airborne sound audible. When it can be heard, the reaction of a listener may be different from their reaction to an airborne sound. Mitigation of groundborne noise where it is excessive may require approaches that are very different from those employed for airborne sound.

Groundborne noise is the result of vibration, but in most contexts the phrase means vibration within the audible frequency range. Although the physics of vibration are directly involved in the study of groundborne noise, the focus on a frequency range higher than is normally the subject of vibration studies means that the underlying mathematics may be applied in a somewhat specialist way.

The source of groundborne noise is frequently mechanical and the transmission media can range from strong rock through softer rocks to soils such as clay, chalk, sand and gravel. Groundborne noise can also travel through water, and soils may have varying degrees of water saturation. The receptor of interest is usually a building, and ultimately a human being listening in the building, although the concern may be an effect on instruments or sensitive

DOI: 10.1201/9781003618775-1

equipment and spaces including sound recording studios. Sources may be situated on or near the ground surface, or be underground, such as in tunnels or caverns. The source excitation may be impulsive, such as piling or breaking in construction work, due to rotating machinery, or moving vehicles. Seismic activity is of course a source of vibration in the ground, and it can be audible, but that aspect of it is of trivial importance compared to its structural effects, which are not considered here.

The study of groundborne noise involves the study of physical phenomena which fall with the broad scope of "Acoustics". In particular the essential vibration results in a wide range of types of wave motion and propagation necessitating solutions to many different wave equations. The dynamic behaviour of the soil or rock to which the source is connected affects the coupling between the source and the medium through which it will radiate vibration, which is the reason for the fact that the comparatively straightforward process in airborne acoustics of quantifying the source strength in terms of its sound power level is much more complex and difficult to determine in the case of groundborne noise.

The first big difference between wave propagation in solid media and in air is the existence of shear stiffness in solid media. This has immensely important consequences and is the main reason for the existence of so many different wave types. In an infinitely large homogeneous medium, the presence of shear stiffness is reasonably straightforward and it results in the creation and propagation, in most circumstances, of shear waves independently of, and in addition to, the pressure waves that are somewhat analogous to sound waves in air. Even when the source does not actually shear the medium, the shear stiffness plays a part in the wave equation for the compression waves. When the medium is not infinite, for example in a halfspace with a surface, a purely compressional wave reaching the surface and being reflected will create reflected shear waves as well as reflected compression waves. If the source is located on the surface, this effect results in the creation of shallow surface waves.

If, below the surface, the propagation medium is not homogenous and either has progressively changing properties with distance from the surface or has a series of layers of different dynamic properties, other types of waves are generated. Waves in layers occur, as do waves at the interface between two media. These waves have the particular property that their propagation speed is dispersive, that is, frequency-dependent.

When waves arrive at a receptor building, the coupling between the soil and the building foundations is one of the most complex problems to solve, and in order to complete the propagation system it is necessary to know the modal response of the building and the extent of amplification due to floor resonances, and, finally, the radiation of airborne sound from vibrating room surfaces.

In airborne acoustics, it is possible to treat sound pressure level and sound intensity almost interchangeably. Because the impedance of air varies little (except in special cases such as radiation from a small point-like source where geometric effects cause impedances to vary from the standard ρc) amplitudes in decibels re 1 picowatt (sound intensity level) and decibel re 20 micropascals (sound pressure level) are numerically similar in most cases. Source strengths are typically stated in terms of sound power levels and, subject to directivity effects, can be used to calculate sound pressure levels in the far field in a very straightforward way.

In contrast to the case for airborne acoustics, with groundborne sound propagation and reception it is necessary to know which of the measurable properties of the vibration is of relevance. Usually, groundborne noise is received as the result of radiation of sound by a vibrating surface, in which case sound pressure level will be proportional to vibration velocity. It will only be indirectly related to power flow or intensity. A given level of vibration power will be associated with high levels of velocity, if the mobility (the inverse of impedance) of the local structure is high, and the reverse is also true. When vibration propagates through a structure, such as a building, it may be observed with varying amplitudes along its path in terms of velocity according to the local mobility, even though the power level may be declining along the path. For this reason, there can be higher velocity amplitudes in more remote locations.

In airborne acoustics, if it is necessary to measure sound intensity, the process is relatively straightforward, involving merely the simultaneous measurement of pressure and velocity (or derivation of velocity from a pair of closely-spaced opposed microphones). To measure intensity or power flow in vibration propagation involves measuring force and velocity, the former of which is difficult to measure. It may be necessary to know in which direction vibration power is flowing, introducing a further complication. For example, in the case of a building with a large footprint affected by vibration from a nearby railway, vibration power may flow into the building in the nearer parts of the building, and out of it through foundations further from the source, causing a damping effect. It can be important to know whether this is occurring, but difficult to establish by measurement.

The units of measurement are basic quantities used in physics and engineering, but they may be unfamiliar to an acoustician used to sound power and sound pressure levels. Parameters that play a minor part in airborne acoustics play a much more significant part in groundborne noise. Forms of wave propagation that are ignored or non-existent in airborne acoustics come to the fore, and their characteristics are much more complex than those of airborne sound waves. The properties of some parameters that play a key role in the propagation of groundborne noise may be very difficult to discover, and, overall, the magnitude of the uncertainties involved in the measurement and prediction of groundborne noise is potentially large.

Sometimes the ground, in the form of soil or rock, is not involved at all, and the transmission path may be all part of one or more structures. The term structureborne noise is used, but in this text it is treated as the same as groundborne noise.

If there is a distinction between groundborne noise and vibration, it is not only that vibration covers frequencies down to fractional Hertz, but also that vibration may be considered in terms of its potential to cause physical damage, or discomfort to human receptors primarily through their sense of touch and motion. These effects are outside the scope of this work even though many principles of vibration need to be discussed in understanding the topic of groundborne noise.

There are many potential sources of groundborne noise, including activities on construction sites, mining and other underground work, fixed machinery installations, and moving vehicles. In the last category, one of the most important sources of groundborne noise is railways, both surface and underground. Often airborne noise from surface railways predominates over groundborne noise (although it is usually present to some extent), but in the case of underground railways and railways in tunnels their audible effect on their surroundings is entirely due to groundborne noise. Many cities of the world have extensive underground railway systems, and mainline railways pass in tunnels near sensitive receptors. Consequently, more work has been done on groundborne railway noise than on any other source, and, as new railways are being planned around the world, the need to predict, assess and mitigate the effect of their groundborne noise is the predominant topic in this field.

Whereas methods of predicting and mitigating airborne noise from railways are now formalised, and established methods and modelling packages are available for the generation of contours and designing necessary mitigation measures, a practitioner seeking to do the same for groundborne noise from railways has to go to a number of sources for information and select an approach that will give results reliable enough to base planning and design decision on them. The advice to be found in available sources varies, and two experts assessing the same railway may produce different results and conclusions, not because of error, but as a result of choosing different sources of advice and interpreting it in different ways. Potential uncertainty is particularly great and is seldom fully addressed.

While the application of airborne acoustical engineering has been formalised to the extent that a practitioner needs only a basic grounding in the underlying physics, in groundborne noise it is necessary to have a deeper understanding of wave motion and vibrational physics. This is not only because of the wide variety of forms of wave motion involved, but also because of the resulting complexity of the mathematics needed to derive even quite basic equations governing the dynamic behaviour of the components and media involved.

In the years since the main authoritative publications became available, the biggest change has been the growth in computing resources. It is now possible to address prediction tasks by numerical simulation in ways which, subject to the accuracy of the assumptions, can greatly reduce uncertainty. The creation of detailed three-dimensional models simulating the dynamic behaviour of rail vehicles, track support systems, support structures, geological conditions and building structures is now possible using desktop computers. While the intensely complex equations that researchers have had to solve to answer even basic questions, such as the behaviour of a plate on an elastic halfspace, provide the opportunity for verification of numerical algorithms, it is no longer necessary to try to calculate, for example, coupling loss factors between the ground and a building, as these can be numerically simulated.

To understand all that follows the reader requires a basic knowledge of acoustics, although, because of the many differences in terminology and units and the substantial extension of the underlying physics, understanding of these is not taken for granted. It is necessary to go into considerable technical detail in many cases. Detailed equations are provided, although a textual explanation is also given.

THE QUANTITIES AND UNITS OF GROUNDBORNE NOISE

Groundborne noise at the receiver is sound, and the units used to quantify it are the same as for all airborne sound. The decibel (dB) is employed along with A-weighting, but, as well as being used to quantify sound pressure levels, the dB scale is also used to express amplitude and ratios of vibration quantities. The same fundamental principles apply in that the dB scale is a ratio between two values. If the quantity involved is dimensional, such as a displacement, velocity or acceleration, both it and the reference for the dB scale are both squared. The same applies to pressure and force. The resulting equation is $L = 10_{\log 10}(A^2/A_{\mathrm{ref}}{}^2)$ where L is the result in dB and A is the measured quantity. Algebraically this is the same as $L = 20_{\log 10}(A/A_{\mathrm{ref}})$.

Many sources published in the USA contain measurements expressed in units of what ANSI (American National Standards Institute) calls the Customary System, largely the same as the Imperial measurement system formerly used in the UK prior to metrication. In particular the units of feet, inches and pounds are used, and the pound is used both for mass and force such that one pound mass is numerically the same as one pound force. Where such results are cited in the following chapters, they have been converted into metric units using the International System of Units (SI) based on the metre-kilogramme-second (mks) system.

However, displacement-based quantities may by expressed in units of metres, millimetres, micrometres or nanometres, and more than one reference level for the decibel scale may be used for the same quantity. Impedances are basically expressed as the ratio of force to velocity. In some cases force per unit area is used which has the SI unit the Pascal, symbol Pa. Mobilities are the reciprocal of impedances. In some publications using metric units, acceleration is expressed in units of g, acceleration due to gravity. To avoid the confusion this can cause, all accelerations in succeeding chapters are in length units of length/time^2. Root-mean-square (r.m.s. or RMS) averages are used, and the system used in airborne noise assessment in which exponential RMS averaging is applied with time constants of 0.125s and 1s ("F" or "S") is used for the assessment of some forms of groundborne noise such as noise from railways.

In the frequency domain, it is often necessary to take phase into account, and the most mathematically convenient way of doing this is by way of complex numbers. Euler's equation $e^{ix} = \cos x + i \sin x$ is extremely useful with $i = \sqrt{-1}$. The first term on the right side is referred to as the real part, and the second as the imaginary part. It is a particularly convenient way of dealing with impedances involving mass-like properties and lossy properties. In wave propagation the part of the equation that accounts for the time and space dependence of the signal and the damping effect uses a wave number with a real part accounting for the former, and an imaginary part accounting for the latter. It is important to know when it is only necessary to take the real part of a result and when both the real and imaginary parts are equally important. The reader unfamiliar with complex numbers will find time spent in familiarisation very beneficial.

Transformation from vibration signals from the time domain to the frequency domain is often required, and some understanding of methods of doing this is needed. The difference between digital filtering and fast Fourier transforms is important, as is windowing in the latter case. It is sometimes necessary to measure impulse responses, and to understand the complexity of frequency analysis of impulses, and the need to use other types of transform. Many of these matters are taken into account by high level applications from MatLab to Python and Excel.

In addition to the differences in the physical and mathematical aspects of groundborne and airborne noise, there are further significant differences when it is necessary to assess the significance of the noise at receptor locations. While a sensitive receptor may be an inanimate instrument or item of equipment, usually it is a human being. As in all cases of sound affecting human beings, the numerical magnitudes and characteristics of sound convey no meaning with regard to the effect on humans, unless linked to the results of studies that have been undertaken in order to correlate physical aspects of sound with human response, whether in the form of annoyance, sleep disturbance, health effects, interference with communication or other psycho-physical effects.

In the case of airborne sound or noise, there is now a substantial body of research into the correlation between the physical properties of the sound and the responses of communities or populations. The response of individual humans is not possible to predict, given the wide distribution of responses that is found in survey results. Only probabilities can be predicted. By contrast with the case for airborne noise, the body of research into the relationship between groundborne noise and human response is very small. What there is has mostly been devoted to the effects of noise from railways, particularly underground railways. There is almost no information relating to other sources of groundborne noise. Practitioners turn to published information relating to airborne noise and seek to adapt it for application to cases of groundborne noise, but there is not always adequate justification for doing this.

Groundborne noise may also affect living species other than human beings. This occurs when, for example, tunnels are being driven under rivers or other bodies of water, and vibration from tunnel-boring machines is transmitted into the water, or when construction work is being carried out in the vicinity of aquaria. This gives rise to a need to understand wave propagation into and through bodies of water, and consideration of criteria for the effects of underwater noise on aquatic species including fish and aquatic mammals.

The same source may cause both airborne and groundborne noise to be perceived at the receptor. An example is where a surface railway passes close to a residential building so that airborne sound enters the building through its weakest part, usually the windows, and groundborne noise enters as vibration through the ground to be radiated as sound inside the building. Because both kinds of sound have the same origin, the assessment of them presents unusual challenges, not least with regard to the choice of assessment criteria.

Where groundborne noise is perceived, there may or may not be vibration perceptible by the tactile senses at the same time. As well as introducing complications in the measurement and assessment of the two effects combined, it may be the case that the person perceiving the effects becomes concerned that the presence of perceptible vibration is an indication of a risk of physical damage to the structure of their building, which in turns adds to an adverse response to the phenomena. In reality, the amplitude of vibration that is accompanied by a risk of structural damage is so much greater than the threshold of perception that human receptors would find the vibration intolerable, and merely hearing and feeling groundborne vibration is not an indication of a significant risk of structural damage.

In the case of proposed new projects. The legal requirement for the completion of an Environmental Impact Assessment (EIA) may mean that vibration and groundborne noise are within the scope of the EIA. Consequently it is necessary to have a system for describing and assigning magnitudes of effect and significance of effects for groundborne noise and vibration. Unlike the case of airborne sound, where criteria for effect magnitudes and significance are now quite well established, it is only in the case of railway projects that the same can be said for groundborne noise and vibration. For other non-railway sources the assessment process may not be straightforward.

Mitigation of groundborne noise and vibration is also much more complicated than is the case for airborne sound, as there are no straightforward equivalents to secondary glazing, noise barriers and the other well-established environmental noise mitigation measures.

Chapter 2

Sources of groundborne noise

Groundborne noise results from vibration at audible frequencies which reaches the ear as sound. The causes of vibration of this kind usually fall into one of the following categories:

- Static machinery and plant
- Moving machinery and plant
- Moving transportation vehicles
- Construction activities
- Blasting, mining and other means of extracting minerals

STATIC MACHINERY AND PLANT

The kinds of machinery that cause vibration include those with rotating components. These may be electrical machines with rotors, turbines or engines with reciprocating components, often involving internal combustion. Examples of the former class are electric motors and generators, and of the latter are reciprocating air compressors and diesel/petrol/gas engines. Vibration can also be caused by plant with no moving parts including electrical transformers where the source is electromagnetic. Fluids flowing in underground pipelines can also cause groundborne noise, for example when there is flow through a pressure reduction value.

MOVING MACHINERY AND PLANT

Moving machinery which is a source of groundborne noise can include a wide variety of items from overhead gantries and tracked dockside cranes to rollers (including vibratory rollers used for ground compaction).

MOVING TRANSPORTATION VEHICLES

This is a large class of sources and includes rail vehicles running on track, on the surface of the ground, or in tunnels or on bridges or other structures.

DOI: 10.1201/9781003618775-2

Moving road vehicles can also be sources of vibration, although vibration from these sources within the audible frequency range is uncommon. Road vehicles running over discontinuities in road paving such as raised ironwork can produce audible vibration.

CONSTRUCTION ACTIVITIES

This is another large class of sources of groundborne noise. The most significant activities are those involving percussive work such as pile driving and rock breaking. Underground tunnelling is an important source of groundborne noise, as are other forms of ground excavation.

BLASTING, MINING AND OTHER MEANS OF EXTRACTING MINERALS

Blasting for the extraction or excavation of rock using explosive charges is primarily a source of vibration perceptible by the tactile sense, although groundborne noise is also involved. Mining and mineral extraction may also involve milling and sawing of rock and percussive work on rock that results in groundborne noise.

GROUNDBORNE NOISE FROM STATIC MACHINES WITH ROTATING COMPONENTS

Static sources of this kind normally have discrete mountings. These are often resilient bearings attached to the machine and to points in the foundation that will often be a concrete slab or box. Amplitudes of vibration at source can, to a limited extent, be calculated from knowledge of the mechanical properties of the rotating components, but these calculations will not normally extend up into the audible frequency range. To establish source strengths in the audible frequency range it is necessary to rely on measurements.

Unlike measurements of airborne source sound power level, where the radiating medium is air and, as long as the measurements are outside the near field, the influence of the radiating medium is readily calculable, the influence of the support structure in the case of measurements of source vibration is not straightforward. There are two ISO Standards devoted to the topic, also implemented by national standard bodies including the BSI, namely ISO 18312-1:2012 and ISO 18312-2:2012 "Mechanical vibration and shock. Measurement of vibration power flow from machines into connected support structures": Part 1 – Direct method, and Part 2 – Indirect method.

Direct method

The ISO direct method begins with the case of a machine directly bolted to a foundation. The vibration power emitted by the machine onto the foundation is defined as the integral over a record length *L* "far greater than the fundamental period of the measured signals" of the time average of the scalar product of the force vector and the velocity vector. These are to be summed over the six degrees of freedom of a mounted machine, namely the three linear or translational degrees (up-down, side-to-side, fore-and-aft) and the three rotational degrees (pitching, yawing and rolling). To resolve this into a frequency spectrum, the standard advises that the real parts of the cross-power spectrum of the force signal and the velocity signal can be obtained from a commercial signal analyser.

ISO 18312-1 then advises on the measurement method. The velocity is measured using a multi-axis transducer (e.g. an accelerometer) on top of the mounting bolt and the force is measured by inserting a force transducer between the machine and the foundation (an alternative method using two force transducers, the second between the mounting bolt and the machine leg, is also given).

Indirect method

The ISO indirect method differs from the direct method in that, instead of inserting a force transducer into the mounting arrangement, the machine is mounted on vibration isolators for which the "blocked transfer dynamic stiffness" is known so that the force can be calculated from the relative velocities of the two ends of each isolator. The term "blocked" means that the dynamic stiffness is determined for the case where the mounting is on an infinitely stiff support.

There are few suppliers of static machines who will respond positively to a request for information on the vibration power of a machine determined according to ISO 18312-1 or 18312-2. In practice, what tends to happen when an engineer is required to predict vibration from a proposed machine installation is that measurements of vibration near an example of the machine installed at another site will be made. What is then required is a method for adjusting those measurements to allow for differences between the two sites.

Assuming that the example and the proposed machines are identical, there are two adjustments necessary. The first is to allow for differences in the mobility of the mounting points in the two cases, and the second is to allow for differences in the propagation characteristics of the structure or ground or both between the machine and the measurement/assessment point.

If the engineer is able to remove the example machine from its mountings, then the mobility of the mounting points can be measured by attaching a multi-axis accelerometer to the foundation and applying a force. There are three methods of applying a force: the first is a blow from an impedance hammer (a hammer with a built-in force transducer); the second is a blow from a hammer with an accelerometer on the hammer head so that the force is the product of the acceleration and the mass of the hammer, and the third is to use an electromagnetic exciter.

If the example machine cannot be removed, then a technique such as that provided in ISO 18312-2 is necessary. The only other approach is to calculate the mobility from knowledge of the properties of the support structure.

Although the driving-point mobilities of a range of types of structure, including beams or plates, are available, calculation of the influence of the ground on which they are supported is intractable, and the practical approach is to create a numerical model from which to output the force and velocity in the time domain, and thereby compute the mobility. Numerical modelling is the discussed in Chapter 10.

Alternative methods of characterising a vibration source are known as the free velocity or blocked force method. Free velocity is the velocity of the source if it were completely unconstrained by a mounting. Blocked force is the force hypothetically exerted on an infinitely rigid support. In theory, therefore, they are ways of expressing the vibrational source strength excluding any influence from its mounting or support. Clearly in reality it is not possible to achieve the mounting conditions assumed in either the free velocity or blocked force case, and indirect methods are needed to measure them in situ.

Blocked force and free velocity are dependent on the source impedance or its inverse, the source mobility such that the free velocity v_{sf} is the product of blocked force and mobility Y_s.

It can be derived by a three-stage measurement process:

1. Measure the coupled velocity of the source on a reference support structure such as a beam.
2. Measure the coupled mobility of the reference support
3. Calculate the blocked force by dividing the results of stage 1 by the results of stage 2.

ISO 20270:2019 specifies a method for the indirect measurement of blocked forces where a source of structureborne sound or vibration is attached to a passive structure. The method specifies a two-stage measurement procedure. First, a passive measurement is made in which frequency response functions (FRF) of the assembled source-receiver structure are measured, and second, vibration is measured with the source operational. The blocked forces are obtained by solving the inverse problem. The Standard explains that

> it is well known that inverse solutions of this type can result in very large errors, particularly if there is inconsistency in the input data. Such errors vary significantly depending on the case and the skill of the operator. Therefore, a means of estimating the uncertainties in the blocked force, through a process called on-board validation, forms an essential part of this measurement procedure.

SOURCES OF VIBRATION ON THE GROUND SURFACE

A point load on an elastic halfspace causes a depression in the surface around it. The displacement equations were first solved by Boussinesq[1] and are:

$$u_x = \frac{P}{4\pi G}\left(\frac{xz}{R^3} - \frac{(1-2\nu)x}{R(R+z)}\right) \tag{2.1}$$

$$u_y = \frac{P}{4\pi G}\left(\frac{yz}{R^3} - \frac{(1-2\nu)y}{R(R+z)}\right) \tag{2.2}$$

$$u_z = \frac{P}{4\pi GR}\left(2(1-\nu) + \frac{z^2}{R^2}\right) \tag{2.3}$$

where P is the load, R is the distance from the load point, G is the shear modulus and ν is Poisson's ratio.

Vibration radiated by a small oscillating source on the surface of an elastic halfspace was studied by Miller and Pursey.[2,3] A notable result of their solution is that for a source (a vertically oscillating force on a small circular disc) on the surface of an isotropic homogeneous halfspace the only wave type which propagates is the Rayleigh wave. On the surface, there is no p-wave or s-wave. However, as soon as the receiving point is moved below the surface, p-waves and s-waves are received. The reason for this is that at grazing incidence the reflection coefficient of a p-wave is −1, but its path is the same as that of the direct wave so that they cancel. Miller and Pursey in their 1955 paper calculated the relative power radiated by such a source by integrating the Rayleigh wave over a cylinder around the source, and the s- and p-waves over a hemisphere, and, in terms of power, the results, for the case where the p-wave speed is three times the s-wave speed, were that 67.4% of the power was radiated as a surface (Rayleigh) wave, 25.8% as a shear wave and 6.9% as a compressional wave. Bearing in mind that velocity amplitude squared is proportional to power divided by impedance and impedance is density multiplied by wave speed, this means that, in velocity amplitude terms, the p-wave percentage should be divided by 3. Given that the Rayleigh wave speed is only slightly less than the shear wave speed (typically about 92%), this means that at any depth below ground the p-wave is likely to be about 16 dB less in amplitude than the Rayleigh or shear wave.

This is important when consideration is being given to the results of field measurements, for example, of transfer mobility involving a hammer blow on the surface and measurements of ground surface vibration a distance from the source.

In reality, it is rare to find a homogeneous isotropic halfspace, and it is common to find that the impedance of the ground increases with depth, and in many locations there is a large impedance change, such as that resulting from the presence of a rockhead below the surface. When the ground impedance increases with depth, this causes Rayleigh waves to become dispersive, and for wave groups to be created whose velocity differs from that of the homogeneous isotropic Rayleigh wave. It is this phenomenon which is used for the computation of ground properties using the MASW method that is discussed in Chapter 9.

The presence of a rockhead or other impedance mismatch is to cause p- and s-waves travelling with a downward component to be reflected, resulting in four reflected waves as discussed in Chapter 3. These waves in turn will be reflected back down from the ground surface, with a p- and an s- reflected wave being generated in each case, subject to the restrictions of the critical angle discussed in Chapter 3. For these reasons, field measurements of the wave radiated across the ground surface by a source such as a hammer blow on the ground surface will pick up p- and s- waves as well as Rayleigh waves. The difficulty arises when it is necessary to calculate the effect of distance, since Rayleigh waves from a point source decay geometrically in proportion to the square root of the distance, while p- and s- waves decay geometrically in proportion to the distance. Furthermore, the effect of loss in the ground material will result in an attenuation in dB per wavelength, and the wavelength of p-waves is much longer than the wavelength of s- and Rayleigh waves.

MOVING SOURCES OF VIBRATION

When a load on an elastic halfspace moves at a speed less than that of the Rayleigh wave speed, a similar depression to the static case given by Boussinesq moves with it (Figure 2.1).

A stationary observer nearby will, therefore, experience a passing wave as the depression progressively deepens and the grows shallow. However, as long as the load is constant, this effect does not propagate.

A similar effect occurs with a group of point loads, such as sequence of wheels as indicated in Figure 2.2.

The vibration experienced by the observer in these cases is of low frequency and sometimes referred to as "quasi-static". It is not a principal source of groundborne noise.

The topic of a moving load on an elastic halfspace was studied by Cole and Huth.[4] The displacements are

$$u_1 = \frac{A}{\pi R(c)}\left[\tan^{-1}\frac{\xi}{\alpha_1 x_2} - \frac{2\alpha_1\alpha_2}{1+\alpha_2^2}\tan^{-1}\frac{\xi}{\alpha_2 x_2}\right] \tag{2.4}$$

and

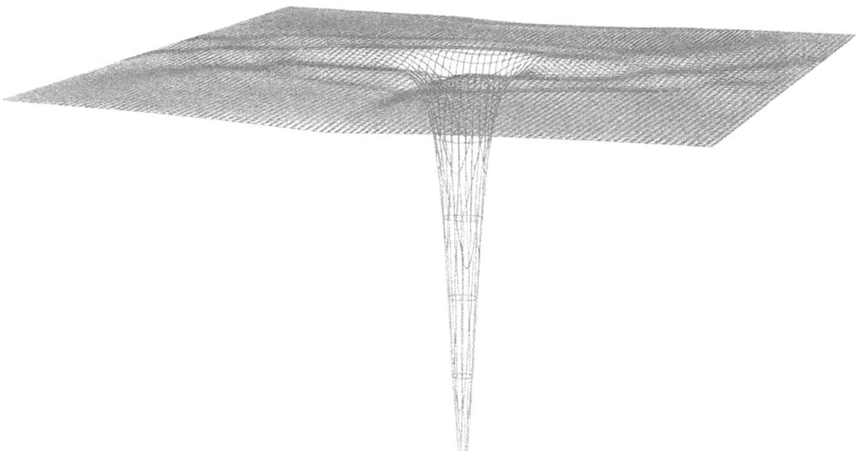

Figure 2.1 Depression in the surface of an elastic halfspace caused by a moving point load.

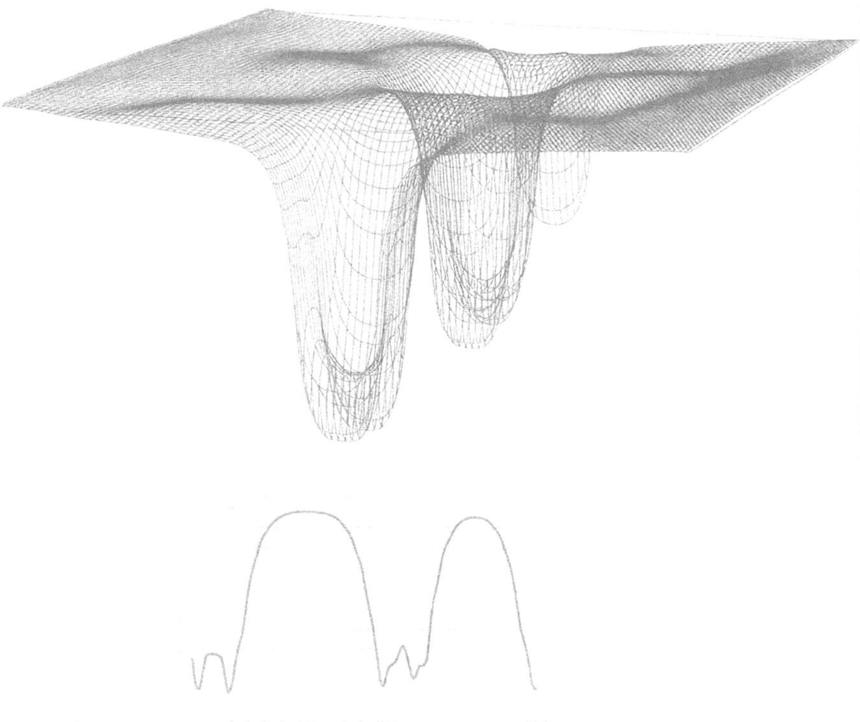

Figure 2.2 Depression in the surface of an elastic halfspace caused by four moving point loads.

$$u_2 = -\frac{\alpha_1 A}{2\pi R(c)}\left[\ln\left(\xi^2 + \alpha_1^2 x_2^2\right) - \frac{2}{1+\alpha_2^2}\ln\left(\xi^2 + \alpha_2^2 x_2^2\right)\right] \tag{2.5}$$

where u_1 and u_2 are horizontal and vertical displacements,

A is $-\frac{(1-\alpha_2^2)P_0}{G}$
$\alpha_1^2 = 1 - c^2/c_1^2$
$\alpha_2^2 = 1 - c^2/c_2^2$
c_2 is the shear wave speed
c is the moving load speed
P_o is the load
$R(c) = \left(1 - \alpha_2^2\right)^2 - 4\alpha_1\alpha_2$
ξ is the moving coordinate $x_1 - ct$
x_1 and x_2 are distances in the horizontal and vertical directions; t is time.

If the speed of the moving load reaches or exceeds the local Rayleigh wave speed, much greater vibration occurs as a "wake", similar to the wake behind a boat, that is generated in the ground surface.

RAILWAYS

The most common case of the moving load is that of a railway, although there is a track between the wheels and the ground. While there is a modified quasi-static effect as the sequence of loads represented by the rail wheels passes along the track, the principal cause of vibration at frequencies within the audible range is the non-smooth nature of rail running surfaces and wheel treads. All rails and wheel treads have minute irregularities which are generally referred to under the heading of "roughness". As wheels run along rails, these irregularities cause forces because the small patch of steel where the wheel tread and rail head are in contact has spring-like properties. Passage of the wheel over the rail irregularities causes the contact spring to be compressed and extended resulting in a force which tends to accelerate the wheel (and its axle) and the rail in opposite directions. The contact patch on the wheel tread is the interface with a network of masses, springs and dampers constituting the wheel, axle, primary suspension, bogie, secondary suspension and vehicle body. From a groundborne noise point of view, the stiffness of the primary suspension is usually low enough to decouple the rest of this network from the wheel, and little error results from considering the system from the contact patch upwards as simply the unsprung mass below the primary suspension. From the contact patch downwards, however, the network of masses, springs and dampers is complex, and it has to be fully taken into account in order to discover how the time-varying force occurring at the contact patch is modified and transmitted by the complete track and its support system.

The characterisation of rail roughness is the subject of ISO 14837-34,[5] and rail roughness spectra are presented in ISO 3095:2005[6] in terms of spectra as a function of wavelength in 1/3 octave bands.

The track support network consists of the rail, rail pads between the rail and a baseplate (if present), resilient support for the baseplate (if present), a sleeper or pair of blocks, and either ballast or a concrete slab in turn on one or more layers of material laid on top of the ground. The ground near the surface is likely to be soil such as clay or sand, below which there may be rock or chalk. Vertically, this involves a sequence of stacks of masses and damped springs, but these stacks are connected by the rail and by the support system below, and those connections involve finite speeds of propagation of bending waves. Moreover, bending wave speeds in beam-like structures are frequency-dependent, so that shorter wavelength/higher frequency waves are propagated at higher speeds.

A railway thus consists of a series of moving point sources of excitation along the track, each supplying a time-varying force into the ground beneath. The forced excitation of the ground beneath is then propagated away into the ground. For a railway on the ground surface (often referred to as "at-grade") the propagation will be mainly by Rayleigh waves and p-waves if the ground is not homogeneous. For underground railways the propagation will be along and around the tunnel walls and invert (if present) and thence as p-waves and shear waves.

It is conventional to regard a railway as a line of incoherent point sources represented by periodic rail supports. However, since the driving force at each support results from wave propagation along the rails and the track, there is a phase relationship between the part of the signal at each support that originates from a specific axle. The input from the succession of axles is a result of wheel/rail roughness on top of quasi-static deflection and these inputs are not normally correlated. However, in some cases, track is installed on a concrete slab which, for vibration mitigation purposes may be supported on resilient bearings. Due to bending wave propagation along the slab there will be a phase relationship between the force beneath successive bearings, and the assumption of a line of incoherent sources may not be valid. The extreme case is when there is full coherence in a line of point sources, that is, at source, they are all in phase.

It is instructive to consider the effect of assuming that a railway is either a line of coherent or incoherent sources.

It is assumed that the railway is on the ground surface, and that it is supported by a series of discrete points separated by space Δx. It is further assumed that vibration from these sources is propagated as a Rayleigh wave and the amplitude is proportional to $1/\sqrt{r_n} e^{-ikr_n}$ where r_n is the distance from the nth point to receiver and k is the wavenumber given by $k = 2\pi f/c_r$ and c_r is the Rayleigh wave speed. To allow for soil loss factor, k is a complex number $k' + ik''$ related to the soil loss factor η according to ISO/TS 14837-32:2015 equation 18 Note 2. For low values of η, k''/k' is approximately $\eta/2$.

The distance r_n from each point source to a receiver is $\sqrt{\left[(n\Delta x)^2 + d^2\right]}$ where d is the normal distance from the track to the receiver.

The received signal at the receiver from each point is $A/\sqrt{r_n} e^{i(\omega t - kr)_n}$ where A is an arbitrary constant. It is not necessary to include the time dependence $e^{i\omega t}$.

If the sources are coherent and the phase of the received signal from each is taken into account the received amplitude signal from the entire line of point sources is

$$V_a = A\left(\frac{1}{\sqrt{r_0}} e^{-ikr_0} + 2\sum_{n=1}^{\infty} 1/\sqrt{r_n}\, e^{-ikr_n}\right) \tag{2.6}$$

and the amplitude will be, using the decibel scale, proportional to $10 \log_{10}|V_a|^2$

If the sources are incoherent phase is neglected and the amplitude is proportional to 10 log V_p where

$$V_p = A\left(\frac{1}{r_0}|e^{-ikr_0}|^2 + 2\sum_{n=1}^{\infty} 1/r_n\, |e^{-ikr_n}|^2\right) \tag{2.7}$$

These two functions are plotted for a point spacing of 0.65 m, and Rayleigh wavelengths of 10 m and 5 m with $\eta = 0.1$ and $\eta = 0.05$ as follows. At 10 m distance, the coherent case gives a result 4–7 dB greater than the incoherent case for the parameters studied. In practice, inhomogeneities in the ground will cause the phase relationship between the received signals from different points to vary and the plotted curve for the coherent case is the envelope representing the case of a perfectly homogeneous ground (Figures 2.3–2.6).

In the vicinity of the track supports, a near field may exist. Near fields involve wave propagation which is emitted at some locations and absorbed at others nearby, setting up closed loops, the energy is which is not propagated to the far field. This may particularly occur when the track is supported on a long floating concrete slab, that is, a slab supported on resilient bearings or on a resilient mat. It happens when the emitting and absorbing locations are vibrating out of phase with each other, and this depends not only on the dynamic characteristics of the bearings, but also on the wave speed, and therefore wavelength, of bending waves in the slab.

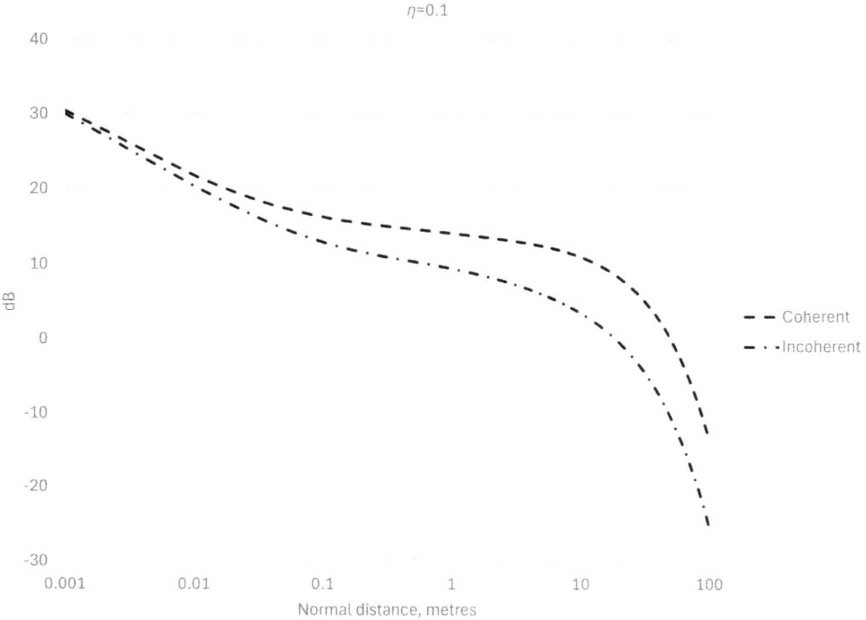

Figure 2.3 Comparison between distance attenuation of a line of coherent and incoherent sources with wavelength of l = 10m and loss factor $\eta = 0.1$.

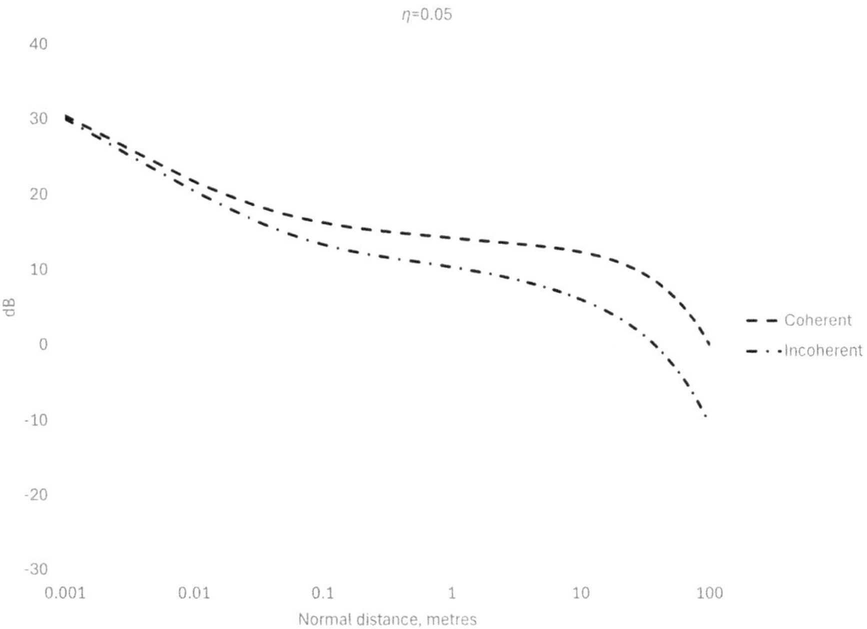

Figure 2.4 Comparison between distance attenuation of a line of coherent and incoherent sources with wavelength of l = 10m and loss factor $\eta = 0.05$.

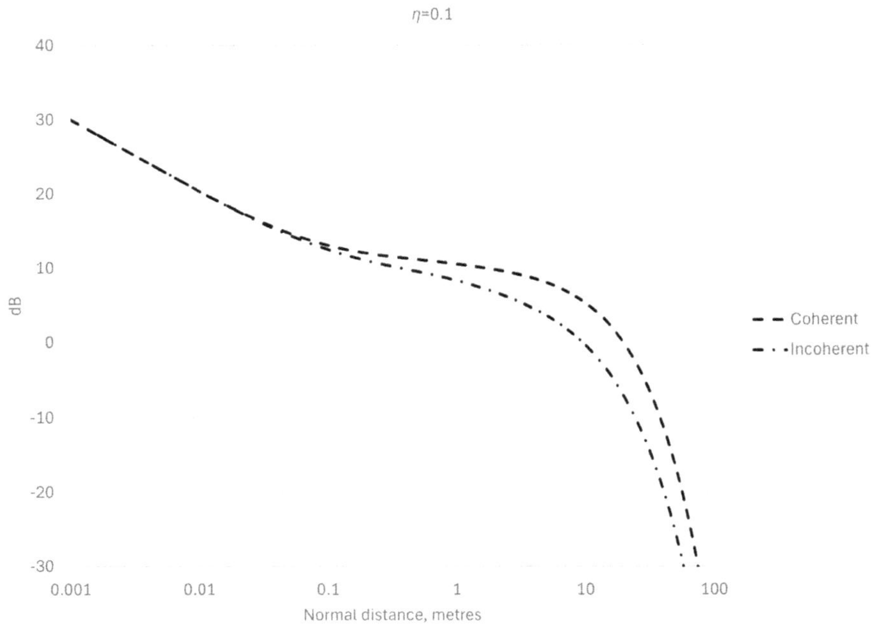

Figure 2.5 Comparison between distance attenuation of a line of coherent and incoherent sources with wavelength of l = 5m and loss factor $\eta = 0.1$.

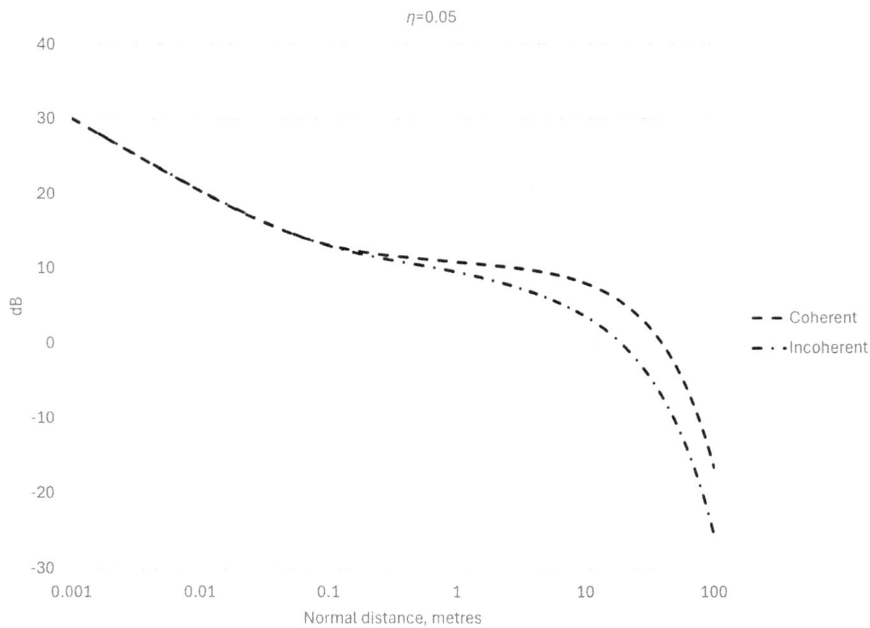

Figure 2.6 Comparison between distance attenuation of a line of coherent and incoherent sources with wavelength of l = 5m and loss factor $\eta = 0.05$.

SOURCES IN TUNNELS

While at-grade railways radiate vibration predominantly in the form of Rayleigh waves as far as receptors at ground level are concerned, railways in tunnels behave differently. Tunnels may be circular in cross section, or rectangular or horseshoe-shaped.

A circular tunnel is cylindrical in form, although there are important differences between a circular underground railway tunnel and the classic "shell" of a cylinder. Nevertheless, it is helpful to consider the vibration behaviour of a cylinder.

An infinitely long cylinder suspended in free space has radial, circumferential and axial modes, to the calculation of which many authors have devoted significant space. Of particular interest are the circumferential modes, of which there is a series beginning with mode zero, which is the cylinder expanding and contracting (sometimes referred to as the breathing mode). Mode 1 is lateral or vertical motion of the entire cylinder, mode 2 is the cylinder ovalling (i.e. so that two bending wavelengths fit around the circumference), and higher modes correspondingly denote the number of bending wavelengths that fit into the circumference. Axial modes relate to wave propagation along the axis of the cylinder. Radial modes result from beam-like behaviour of the cylinder.

The effect of the system of cylinder modes is primarily to determine the driving point impedance of the wall of the cylinder, which affects the transmission of vibrational power from a railway in the tunnel into the tunnel wall, but they also affect the way the vibration, after entering the tunnel wall, is distributed around its circumference and propagated along its length.

In reality a railway tunnel is not a plain cylinder, because it nearly always has an invert, that is, a layer of concrete at the base, and it is embedded in the elastic medium of the surrounding soil or rock. (Tunnels in rock are often not cylindrical in section). Furthermore, the walls of a railway tunnel may not be continuous solid concrete, but formed of segments, either of concrete or cast-iron, bolted together. The extent to which moment in tunnel lining segments is interconnected so as to cause it to behave as a cylinder depends on the nature of the joints between the segments. The interface between the tunnel walls and the surrounding soil may also be variable. Frequently the space around tunnel lining segments is injected with grout of variable thickness, and there can be voids between the tunnel extrados and the soil. These features affect both the frequencies of the tunnel modes, and also the transmission around the tunnel wall of vibration from the excitation points at the base of the tunnel.

These effects are apparent when the vibration of a railway tunnel embedded in soil is simulated numerically, and a cross section shows that vibration is concentrated near the base of the tunnel in the region of the track supports and not distributed equally around the circumference. Such a simulation also shows the relative extent to which shear and p-waves are radiated into the soil, this being important from the point of view of considering attenuation with distance from the tunnel. Shear waves are attenuated for geometric reasons at the same rate as p-waves, but the effect of soil-loss factor is such as to give an attenuation in dB per wavelength, and shear wavelengths are much shorter than p-wavelengths.

Rectangular tunnels exhibit a basically similar behaviour in that there are the equivalent of circumferential, radial and axial modes but for the same size of tunnel they will occur at different frequencies. When a bending wave propagates in a plate and meets a right-angled bend, even when the plate thicknesses are the same on both sides of the bend, reflection occurs with a reflection coefficient of $1/\sqrt{2}$ (i.e. -3 dB). This means that, for excitation at the base of the tunnel, there is less transmission to the walls of a rectangular tunnel than there is in the case of a cylindrical tunnel.

Because the source of vibration in an underground railway tunnel tends to be concentrated near the base of the tunnel, this means that the tunnel itself is an obstruction in the path of vibration radiation in upward directions and it can act as a "noise barrier" in the path of groundborne noise radiated from the base of the tunnel to receptor locations above the tunnel. Depending on the extent to which vibration is transmitted around the circumference of the tunnel, it is possible to find that measured ground surface vibration is lower directly above the tunnel, increasing with the distance of the centreline, before decaying beyond a point several metres from the centreline.

For these reasons, when carrying out predictions of vibration from an underground railway tunnel, determination of the source strength and its location is not a simple exercise.

Force density

A method of quantifying the source strength of an existing railway is to use the property known as force density. This is a spectrum in units of $N/m^{0.5}$. This is a property that cannot be measured directly but is derived from (a) measurements of line-source transfer mobility and (b) vibration from a train passage. The line source transfer mobility is in units of $(m/s)/(N/m^{0.5})$. Measurement of the line source transfer mobility involves first measuring the point source transfer mobility, for example, by impacting a source location with a hammer or dropping a weight on to a load cell and taking the frequency transforms of the input force and of the velocity at the receiving location or a sequence of locations at increasing distances. This is then repeated for a series of source locations along the line being measured, and the set of point source mobilities are combined into a line source mobility estimate by numerical integration using Simpson's rule. Assuming that there is an equal distance h between and even number n of point sources, and the square of the mobility for each is $f(x)$ where $x = nh$ then the integral is

$$h/3[f(0) + 4f(h) + 2f(2h) + 4f(3h) + \cdots + \; + 4f((n-1)h) + f(nh)] + \epsilon(h) \tag{2.8}$$

where $\epsilon(h)$ is an error term that for this application is small enough that it can be neglected.

Care is required interpreting the scales used to express force density, which may be in decibels re 1 $lb/ft^{0.5}$, or decibels re 1 $N/m^{0.5}$. 1 $lb/ft^{0.5}$ is 8.057103 $N/m^{0.5}$. This means 0 dB re 1 $lb/ft^{0.5}$ is 8.057103 $N/m^{0.5}$, and expressing this as dB re 1 $N/m^{0.5}$ gives a result 18.124 dB higher.

Transfer mobility may be in (a) dB re 10^{-6} $in/s/\left(lb/ft^{0.5}\right)$ or (b) dB re $10^{-8} m/s/\left(N/m^{0.5}\right)$ or (c) dB re $10^{-9} m/s/\left(N/m^{0.5}\right)$. To convert (a) to (b), subtract 10 dB. To convert (a) to (c), add 10 dB.

Typical force densities are given in the FTA report 0123.[7] An earlier Transportation Research Record report[8] also gives a number of force density results. More recently an Austrian paper[9] gives a set of Force densities for seven different types of train. Force density results for 117 passenger train passages (77–100 km/h) are reported in the Silvarstar Deliverable D1 report.[10]

Liu et al.[11] found that the force density measured for one car in a train was nearly the same as that for the whole train.

An important consideration is that force density results are not for blocked-force, that is, independent of the impedance of the supporting structure. The force density level depends on both the force generated at the wheel/rail interface and the dynamic characteristics of the supporting structure, including the track, subgrade and soil. To make use of force density in predictions, it is necessary first to measure or compute the line transfer mobility for the site in question, and that will relate force as it is manifested at the source point – for example, the load cell impacted by the hammer blow or falling mass – to velocity in the field. If a force-density spectrum for a train at another site is then used to predict likely ground velocity, then to minimise uncertainty associated with the dynamic characteristics of the source medium, in the case for which the force density spectrum was measured, it is necessary to estimate the difference between the support impedances in the two cases. That is, it is necessary to answer the question "what would the force density of the source train have been if it had been running on the same support system as will occur in the case to be predicted?".

It can be determined numerically using a finite element or finite difference model.

CONSTRUCTION SOURCES

Sources of groundborne noise in construction mainly involve excavation of one form or another, although structureborne noise can result when building work is carried out on a source structure in contact with a receiving structure. In the former category the largest source is associated with tunnel excavation, particularly tunnelling using a tunnel-boring machine.

Tunnel boring machines

Tunnel boring machines are used to excavate tunnels in all kinds of soil and rock, from clay to sedimentary and even igneous rock, including granite. They result in varying degrees of groundborne noise and vibration for receptors within range, and it is usually necessary to carry out predictions of likely levels.

A tunnel boring machine, abbreviated TBM, is a cylindrical framework at the front of which is a rotating cutter head, behind which is machinery to handle tunnel-lining segments and to remove the arisings from tunnelling by conveyor or a pipe system. Some machines use slurry at the face, and the arisings are pumped back in the slurry, for separation above ground. There is usually a system for achieving earth pressure balance to minimise ground settlement due to face loss. The most significant groundborne noise is caused by the cutting action of the cutter head, and there is some caused by machinery which can be heard in receptor buildings during periods when the cutter head is idle. TBMs come in all sizes from a few metres to as much as 20 m in diameter (Figure 2.7).

Given the uncertainties involved in attempting to predict the vibration caused by a TBM from first principles, the first step in a prediction process is normally to obtain measurements from the operation of an actual TBM. The difficulty is that it is quite uncommon to find results for the same geology

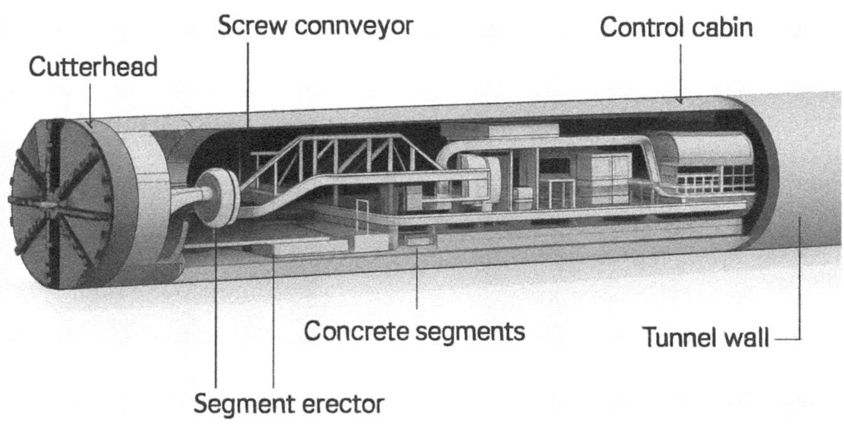

Figure 2.7 Schematic view of tunnel boring machine.

Source: Courtesy of Herrenknecht AG.

and TBM diameter, and the data obtained will have to be adapted to take account of differences between the machine measured and the proposed system. The most important parameters which vary are tunnel diameter, soil or rock properties, cutter rotational speed and face pressure.

The effect of tunnel diameter, for the same geology, is a function of the relative face areas, and the power absorbed by the cutting process. If the powers of the measured and proposed TBMs are known as W_1 and W_2, the total power emitted as groundborne noise and vibration will vary as $10_{\log 10}(W_2/W_1)$. However, the cutter-head diameter is also important, particularly when it is comparable with distances to receptors, because it will influence the radiation pattern. A TBM cutter face may be regarded as a disc across which is distributed an array of incoherent point sources, which simplifies the radiation pattern somewhat, but it will still be the case that the effect of distance will be small until the distance is significantly greater than the tunnel diameter. There is a strong directivity effect, which not only influences the received level of groundborne noise for the TBM at any particular point in its progress, but which also affects the rate of rise and fall in noise level as the machine approaches, passes under the receptor and departs.

The vibration generated at the cutter face is somewhat directional. If vertical vibration is measured in the ground, it will tend to be higher some distance ahead of the face than directly above it. The duration of groundborne noise is obviously dependent on the rate of progress of the TBM, which in turn will be dependent on the geology and performance characteristics of the machine.

The effect of geology is also extremely important, in that the groundborne noise of a TBM in clay is very much less than a TBM in granite, for example, and in mixed soils such as sand and gravel, or boulder clay, there can be local variations, for example, if a boulder is encountered and is broken up by the cutter head. Unlike rocks, soils do not have to be fractured, with the result that the groundborne noise is very different in these two cases. Under the heading of "rock" are to be found three main types, namely igneous, sedimentary and metamorphic rocks. Igneous rocks result from the solidification of molten volcanic magma, and a frequently encountered example is granite. Tunnelling for the Singapore Mass Rapid Transit (MTR) was carried out in granite using a TBM. However, it is important that variations in the nature of rock of this kind with depth are taken into account, because the surface of the granite will have undergone a weathering process that tends to decompose the granite. In fact, the soil in parts of Singapore is "completely decomposed granite" which is granular, soil-like material. Between a stratum of competent granite and the overburden there is likely to be a progressive succession of layers of varying degrees of weathering, which significantly different properties. Sedimentary rocks include sandstone, limestone, mudstone and chalk and these are also capable of excavation using TBMs. Chalk in particular is likely to have layers of weathered material toward the top of a chalk layer and this may be so decomposed that the chalk is more like soil than rock. Sedimentary rocks are softer than igneous rocks.

Metamorphic rocks, such as gneiss and schist are formed from igneous or sedimentary rocks that have undergone alteration by way of heat and pressure or chemical reactions deep within the earth's crust.

Soils include cohesive soils such as clay and silt and non-cohesive soils such as sand and gravel. Peat is also encountered sometimes in tunnelling.

The differences between soils and rocks are great enough for it to be inadvisable to attempt to adapt measurements of groundborne noise and vibration from a TBM in soil to appropriate characteristics in rock.

A first step, in the case of rock, in comparing the likely potential for generating groundborne noise during cutting by TBM is to compare the energy of fracture of the rock. Energy of fracture is the energy required to break intermolecular bonds and form new surface area and is measured in units of Joules per square metre. It is related to fracture toughness in $Pa/m^{0.5}$ Following fracture, the released energy is dissipated as vibration.

As an example, the fracture toughness of granite is approximately 2.2 times that of sandstone. This will in large part account for the difference in power absorbed by a TBM in granite and a TBM in sandstone.

The properties of the medium in which the TBM is operating not only affect the source strength, but also the propagation characteristics. While the effect of geometric spreading of vibration from the cutter head will be similar for two machines of the same face diameter, the loss with distance due to material damping will be different. Competent rock has a loss factor η of no more than 0.05, while a cohesive soil may be between 0.05 and 0.1 and a non-cohesive soil higher than 0.1. The equations for the propagation of vibration in soil and rock are discussed in Chapter 3. The topic of coupling loss factor and building response, which must be taken into account when predicting TBM groundborne noise in buildings is discussed in Chapter 4.

Other sources of groundborne noise in construction

While tunnel boring machines are among the largest and most powerful sources of groundborne noise in construction, noise can also be caused by activities such as secant piling, where a line of primary piles is first constructed and then secondary piles are inserted between them with the augur cutting into the primary piles to create an interlocking wall. Groundborne noise is caused by the augur cutting the primary piles. An alternative to the secant pile wall is a diaphragm wall, in the construction of which excavation takes place, using a machine with a rotating cutter head (a drum with teeth). Diaphragm walls may be constructed in geological conditions where soil overlies a harder stratum such as chalk or rock. When the cutter head reaches the harder stratum, it will cut into it for some distance so that the wall is then cast by pouring concrete around a reinforcement cage inserted into the deep trench that has been excavated by the cutter head.

The action of the cutter head, sometimes called a hydromill, is a source of groundborne noise.

Excavation in rock may be carried out using a machine often referred to as a road header. This is also a machine with a rotating toothed drum on the end of a boom, used for excavation of the ground including rock, causing groundborne noise. Hydraulic breakers or hammers, "peckers", may be used to excavate rock, with consequent groundborne noise from the percussive activity on the rock. Rock saws can be used, somewhat like chainsaws. Even when low noise techniques are used such as bursting using expanding grout, it is necessary to drill holes in the rock to insert the grout, and the rock drills involved cause groundborne noise.

There is very little published information about the source strength of groundborne noise generators such as these. Prasad[12] reported the results of measurements of vibration from the construction of a diaphragm wall in boulder clay above limestone, involving the use both of a hydromill and a diaphragm wall grab. For the frequency range up to 100 Hz, the results are summarised in Figure 2.8.

Menon et al.[13] published measured A-weighted groundborne noise levels in sandstone from use of a TBM and a hydraulic hammer for cross-passage construction, which, for the TBM showed an almost linear decay from approximately 71 dBA at 20 m to 35 dBA at 287 m, and for "intense hammering" 50 dBA at 55 m decaying to 42 dBA at 150 m.

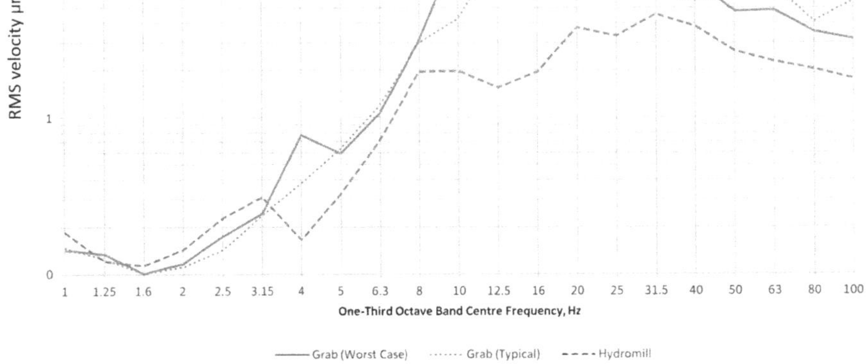

Figure 2.8 Diaphragm walling source spectra at 50 m.

Wilkinson Murray presented a summary of groundborne noise levels from road headers.[14]

While most published data on vibration from construction activities are given in terms of vibration and frequently present only peak particle velocity figures, there are some groundborne noise data for tunnelling operations available, albeit only in dB(A) terms, in TRL Report 429.[15]

MICROSEISMS

Microseisms are faint earth tremors caused by natural phenomena. Seismograms can be found to pick up background noise. These may be Rayleigh waves caused by water waves in the oceans and lakes, wind in the trees and by small seismic sources in the Earth's mantle. Their spectrum may extend into the audible range so that they are potential sources of groundborne noise, although seldom of an amplitude that makes them significant.

NOTES

1. Boussinesq, J. *Application des potentials a l'etude de l'equilibre et des mouvements des solides elastiques*. Ganther-Villars, Paris, 1885.
2. Miller, G.F., and Pursey, H. *The field and radiation impedance of mechanical radiators on the free surface of a semi-infinite isotropic solid*. The National Physical Laboratory, Teddington, Middlesex, 1953.
3. Miller, G.F., and Pursey, H. *On the partition of energy between elastic waves in a semi-infinite solid*. The National Physical Laboratory, Teddington, Middlesex, 1955.
4. Cole, J., and Huth, J. "Stresses produced in a half plane by moving loads." *ASME Journal of Applied Mechanics*, 25, 433–436, 1958.
5. ISO/DTS 14837-34. *Mechanical vibration—Groundborne noise and vibration arising from rail systems—Part 34: Characterizing irregularity of the running surfaces with respect to vibration excitation*. International Organization for Standardization, Geneva, 2024.
6. ISO 3095-2005. *Railway applications—Acoustics—Measurement of noise emitted by railbound vehicles*. International Organization for Standardization, Geneva, 2005.
7. Volpe, John, A. "Transit noise and vibration impact assessment manual." FTA Report No. 0123, Federal Transit Administration, National Transportation Systems Center, United States, 2018, 319. https://www.transit.dot.gov/sites/fta.dot.gov/files/docs/research-innovation/118131/transit-noise-and-vibration-impact-assessment-manual-fta-report-no-0123_0.pdf
8. Nelson, J.T., and Saurenman, H.J. "A prediction procedure for rail transportation ground-borne noise and vibration." *Transportation Research Record*, 1143, 26–35, August 1988.

9. Tappauf, B., Alten, K., Legenstein, M., Ofner, M., and Flesch. R. "Experimentally determined force density spectra for admittance-based vibration predictions along railways." *Applied Science*, 14, 2557, 2024. https://doi.org/10.3390/app14062557
10. Degrande, G., Lombaert, G., Ntotsios, E., Thompson, D., Nélain, B., Bouvet, P., Grabau, S., Blaul, J., and Nuber, A. "State-of-the-art and concept of the vibration prediction tool." SILVARSTAR Deliverable D1.1, 28 May 2021.
11. Liu, W., Wu, Z., Li, C., and Xu, L. "Prediction of ground-borne vibration induced by a moving underground train based on excitation experiments." *Journal of Sound and Vibration*, 523, 116728, 14 April 2022.
12. Prasad, D. "Assessment of low vibration techniques in the construction of a diaphragm wall using a hydromill." *Proceedings of the Institute of Acoustics*, 36(Pt.3), 2014.
13. Menon, V.A., Miller, A., Hanson, D., and Parnell, J. "Advanced analysis of ground-borne vibration noise impacts for underground railway tunnel construction." *Proceedings of Acoustics*, Gold Coast, Australia, 6–8 November 2024.
14. Wilkinson Murray Pty. Ltd. "Australian Acoustical society technical meeting – Tunnelling noise and vibration management." Wilkinson Murray, 2003.
15. Hiller, D.M., and Crabb, G.I. "Groundborne vibration caused by mechanised construction works." TRL Report 429, Transport Research Laboratory, 2000.

Chapter 3

Wave propagation

The term "wave equation" applies to the formulation of the way a small region of an elastic medium responds to a disturbance. It is convenient to treat the disturbance as the physical displacement of a particle (effectively just a notional point in the medium) from its rest position.

When a particle in an elastic medium is displaced from its rest position, strain occurs around it, and that strain results in stress. Strain is a dimensionless phenomenon representing the change in the length of a dimension as a ratio of the changed dimension to the unchanged dimension. Stress is the internal force per unit area that results from the deformation described by the strain as a result of the elasticity of the medium. Force tends to accelerate a particle according to the principle force = mass × acceleration or acceleration = force/mass. The combined result of these effects is that, when a particle is disturbed from its rest position, it will tend to be accelerated back towards its origin, but that acceleration will impart velocity to it, as a result of which it will overshoot its origin, changing the sign of the stress. What follows is oscillation, both in time and space. It is this oscillation that gives rise to wave motion.

Particle displacement can be stated in length units, and it is traditional to use the Greek letters ξ, η and ζ to refer to the displacement in the three directions x, y and z. If two adjacent particles are displaced by unequal amounts, the medium between them will be either compressed or stretched. The amount of such compression is stated as $\partial\xi/\partial x$. The symbol ∂ represents the difference between the positions of an adjacent pair of particles. This ratio is dimensionless, as both the numerator and denominator are units of length. Looking at one of the two particles, the consequence of the compression will be a force on the particle, as a result of compressing the spring-like elastic medium. A starting point is to look at the one-dimensional case, assuming a compressive medium confined in a tube so it cannot expand or contract laterally, only longitudinally in the x direction. If that elasticity is expressed as a compression or constrained modulus D, in units of Newtons per square metre, the force will be $D\partial\xi/\partial x$, also in units of N/m^2. What effect this force will have on the particle will

DOI: 10.1201/9781003618775-3

depend on what has happened on its other side, that is, $\partial\xi/\partial x$, where $\partial\xi$ is this time the displacement of that particle on the other side. The net force on the particle will be the difference between the force due to compression on one side and the compression on the other side. This is written as $\partial^2\xi/\partial x^2$, with dimensions 1/m. The net force on the particle will be $D\partial^2\xi/\partial x^2$. From fundamental mechanics, force = mass × acceleration and, therefore, acceleration = force/mass, so to complete the equation, the acceleration of this particle will be $D/\rho\,\partial^2\xi/\partial x^2$ where ρ is the mass density in units of kilogrammes per cubic metre which can also be written as Ns^2/m^4. This gives an answer in units of N/m^2 divided by Ns^2/m^4 to give m/s^2 as follows

$$D\frac{\partial^2\xi}{\partial x^2} = \rho\frac{\partial^2\xi}{\partial t^2} \tag{3.1}$$

Rearranging this gives $\partial x/\partial t = \sqrt{\frac{D}{\rho}}$, the speed at which the pattern of displacements of particles travels in the x direction, that is, the speed of sound.

Moving from the one-dimensional case to the three-dimensional case, assuming a medium with zero shear modulus, such as a gas, it is necessary to include the effect of lateral contraction or expansion, since compressing a gas results in a force acting in all three directions.

$$D\left(\frac{\partial^2\xi}{\partial x^2} + \frac{\partial^2\eta}{\partial x\partial y} + \frac{\partial^2\zeta}{\partial x\partial z}\right) = \rho\frac{\partial^2\xi}{\partial t^2} \tag{3.2}$$

While the absence of shear modulus applies to fluids, solids have shear moduli, and they modify the wave equation in two ways. Firstly, when the compressible medium has a non-zero shear modulus, compressing it does not result in a lateral force the same as the force in the direction of compression. Compressing a material with a shear modulus causes it to bulge sideways, but the lateral extension is less than the longitudinal contraction, and the ratio of the two is called the Poisson's ratio, given here the symbol σ. Poisson's ratio determines the lateral force as a proportion of the force that would occur in a zero-shear medium, and that proportion is $\sigma/(1-\sigma)$. Equation 3.2 is now modified to

$$D\left[\frac{\partial^2\xi}{\partial x^2} + \frac{\sigma}{(1-\sigma)}\left(\frac{\partial^2\eta}{\partial x\partial y} + \frac{\partial^2\zeta}{\partial x\partial z}\right)\right] = \rho\frac{\partial^2\xi}{\partial t^2} \tag{3.3}$$

$D\sigma/(1-\sigma)$ is equivalent to $D-2G$.

Poisson's ratio determines the value of the shear modulus, G, of an elastic material, according to the equation

$$G = D\frac{1-2\sigma}{2(1-\sigma)} \tag{3.4}$$

The second effect that results when a medium has a non-zero shear modulus is that, in addition to the compression waves just described (also called dilatational waves or p-waves), there also exist shear waves. In an infinite homogenous isotropic medium, shear waves exist independently of the p-waves, and the derivation of the shear wave equation is similar to that for p-waves, except that instead of compression causing a force, it is the sliding of two particles relative to each other that results in shear force. Thus Equation 3.1 becomes

$$G\left(\frac{\partial^2 \xi}{\partial y^2} + \frac{\partial^2 \xi}{\partial z^2} + \frac{\partial^2 \eta}{\partial x \partial y} + \frac{\partial^2 \zeta}{\partial x \partial z}\right) = \rho \frac{\partial^2 \xi}{\partial t^2} \tag{3.5}$$

Unlike p-waves, in which the direction of particle motion is the same as the direction of wave propagation, in the case of shear waves, or s-waves, the particle motion is at right angles to the direction of propagation. Furthermore, it can be in any direction in the *yz* plane if the wave is travelling in the *x*-direction. If the *z*-direction is up or down, waves involving particle motion in the *z*-direction are referred to in seismology as SV waves, and in the *y*-direction as SH waves.

Despite the basic independence (in an isotropic medium) of shear waves, they usually co-exist with p-waves, so that the complete equation (ignoring damping, i.e., loss-free) is

$$G\left(\frac{\partial^2 \xi}{\partial y^2} + \frac{\partial^2 \xi}{\partial z^2} + \frac{\partial^2 \eta}{\partial x \partial y} + \frac{\partial^2 \zeta}{\partial x \partial z}\right) + D\left[\frac{\partial^2 \xi}{\partial x^2} + \frac{\sigma}{(1-\sigma)}\left(\frac{\partial^2 \eta}{\partial x \partial y} + \frac{\partial^2 \zeta}{\partial x \partial z}\right)\right] = \rho \frac{\partial^2 \xi}{\partial t^2} \tag{3.6}$$

for the *x* direction, with corresponding versions for the *y* and *z* directions. All three can be combined in a single vector equation.

Frequently, the elasticity of a medium is stated in terms of Young's modulus, given the symbol *E*. An important fundamental principle is that if any two of the four constants *D*, *E*, *G* and σ are known, the other two can be derived from them. The relationships are

$$E = 2G(1 + \sigma) \tag{3.7}$$

$$\sigma = \frac{2G - D}{2(G - D)} \tag{3.8}$$

$$\frac{G}{D} = \frac{1 - 2\sigma}{2(1 - \sigma)} \tag{3.9}$$

$$\frac{E}{D} = \frac{(1+\sigma)(1-2\sigma)}{(1-\sigma)} \tag{3.10}$$

It follows from these relationships that Equation 3.6 can be written in a number of ways, and some texts may write it in terms of Lamé constants λ and μ. Of these μ is the same as G; λ is $D\sigma/(1-\sigma)$ or $D-2G$ and $\lambda + 2\mu = D$. Equation 3.6 is then written as

$$\mu\left(\frac{\partial^2 \xi}{\partial x^2} + \frac{\partial^2 \xi}{\partial y^2} + \frac{\partial^2 \xi}{\partial z^2}\right) + (\lambda+\mu)\left(\frac{\partial^2 \xi}{\partial x^2} + \frac{\partial^2 \eta}{\partial x \partial y} + \frac{\partial^2 \zeta}{\partial x \partial z}\right) = \rho \frac{\partial^2 \xi}{\partial t^2} \tag{3.11}$$

In the prediction of groundborne noise propagation, knowledge of the values of G and D for the medium concerned is necessary because, together with the mass density ρ, they determine the shear and p-wave speeds. The wave speeds are important when it comes to predicting the attenuation, due to material damping, which is proportional to distance measured in wavelengths rather than distance units. Wavelength is c/f, where c is the wave speed and f is the frequency in Hz.

It is important to be aware that other moduli besides D and G are in use and they should not be confused. The constrained modulus D is sometimes given the symbol M, and D and M are the same, and represent compression of a volume under load when the sides are constrained. However, the bulk modulus, $K = D - 4G/3$, is the ratio of pressure increase to decrease in volume, when all surfaces are compressed, and Young's modulus $E = 2G(1+\sigma)$ is the compressive stiffness of a volume when the sides are free. When the topic is groundborne noise, each of the modulus symbols should be followed by the subscript *max* in order to signify that they are small-strain values, as the moduli for most elastic materials are decreases at large strains. A useful set of interrelationships is

$$D = E(1-\sigma)/((1+\sigma)(1-2\,\sigma))$$

$$D = G(4G-E)/(3G-E)$$

$$G/D = (1-2\sigma)/(2(1-\sigma))$$

$$\sigma = (2G-D)/(2G-2D)$$

DAMPING

The wave equation set out above does not take damping or loss factor into account, and additional terms are needed to account for it. A non-zero loss factor causes the force resulting from compression or shear strain to lag behind the strain. The amount of the lag can be expressed as an angle, in radians, where a lag of a half wave cycle is π radians. This leads to a convenient way of considering loss factors by expressing the values of D and G as complex numbers. Complex numbers have two parts, real and imaginary, and the essential feature of them is that the ratio of the imaginary part (given the prefix i) to the real part is the loss factor, η. The angle of lag is the arctangent of the imaginary part divided by the real part. This means that, from a laboratory measurement of stress and strain on a test sample, the loss factor can be calculated from the relationship between the two curves, when the strain oscillates at the test frequency.

Using a complex modulus leads to a way of calculating the loss due to the damping resulting from the value of η. The method makes use of Euler's formula which shows that the exponent of an imaginary number e^{ia} can be written as $\cos a + i \sin a$. Since a wave of a single frequency propagating through the ground has an amplitude which, at any given time, varies as $\cos kx$ where k is the wavenumber $2\pi f/c = 2\pi/\lambda$, an interesting result is obtained if a is a complex number (which it will be if c is complex) because the modulus D or G in $c = \sqrt{\frac{D}{\rho}}$ or $v = \sqrt{\frac{G}{\rho}}$ is complex. The square of i is -1 so $e^{i(\alpha + i\beta)}$ becomes $(\cos \alpha + i \sin \alpha)e^{-\beta}$. The term $e^{-\beta}$ does not oscillate but simply reduces the amplitude of the oscillating term as the distance increases.

If β/kx is called the distance attenuation factor, it can be calculated from η using the equation

$$\beta = kx\sqrt{\frac{\sqrt{1 + \eta^2} - 1}{2(1 + \eta^2)}} \tag{3.12}$$

For low values of η, β/kx is approximately equal to $\eta/2$.

The vibration which leads to the occurrence of groundborne noise is conveniently expressed using the decibel scale, and in terms of vibration velocity amplitude the dB scale is given by $L_v = 20 \log_{10}(V/V_{\text{ref}})$, where V_{ref} is 1 nanometre/s. There is a simple relationship between a naperian logarithm (which is what $-\beta$ is in $e^{-\beta}$) and a common logarithm (log to the base 10), and the former is converted to the latter by multiplying by $\log_{10} e$. That is, multiplying $\beta = kx$ by 0.43429 means that the attenuation due to distance in decibels is approximately $4.34\ \omega\eta x/c\,$dB.

Damping can take several forms, including viscous, friction and hysteretic damping. In the time domain it is possible to create a frequency dependent (including a frequency-independent) damping term with any desired spectrum shape, using the relaxation principle of Boltzmann,[1] where

$$s(t) = D_1\,\epsilon(t) - \int_0^{\infty} \epsilon(t - \Delta t)\phi(\Delta t)d(\Delta t) \tag{3.13}$$

where $\phi(\Delta t) = \frac{D2}{\tau}e^{-\Delta t/\tau}$ is an after-effect function, D_2 is a constant and τ is a relaxation time. D_1 is a modulus, $s(t)$ is stress and $\epsilon(t)$ is strain.

The loss factor η is given by

$$\frac{D_2\omega\tau/(\omega^2\tau^2+1)}{D_1-D_2/(\omega^2\tau^2+1)} \tag{3.14}$$

By combining several after-effect functions with different values of D_2 and τ, any relationship between loss factor and frequency may be represented. Note that the real part of the modulus, the denominator of Equation 3.14, is reduced in magnitude, with the result that the value of the modulus is reduced by the inclusion of the relaxation terms. Depending on the choice of the constants and relaxation times, the stiffness of a resilient element will be frequency dependent, and the value of D_1 must be adjusted at the same time that D_2 and τ are selected to give the required dynamic stiffness.

While data on the small-strain values of G and σ are not difficult to obtain, information about the loss factor η is rarely measured or published. The measurement of η is not straightforward. It can be measured in the laboratory by a variety of means, or in the field. Laboratory measurements will yield a relatively "pure" η, that is, loss solely due to the inherent properties of the medium itself. Field measurements, such as may be obtained using two or more boreholes and extracting the excess attenuation, are subject to two confounding influences. The first is that, unless the boreholes are widely separated, there will be near-field effects which result in attenuation for geometric reasons unrelated to the loss factor, and to calculate these and remove them from the results is a complex exercise. Secondly, if measurements are made over a large enough distance to minimise near field effects, unless the soil is perfectly homogeneous, there will be losses due to reflection and refraction phenomena, and wave conversion, at the interface of layers with different velocities, and due to scattering from inhomogeneities.

On the other hand, in reality, these additional effects do occur, so carrying out a prediction using "pure" loss factor will in such cases underestimate the excess, non-geometric, attenuation.

There are several types of laboratory tests which can be used to measure loss factor, and one of the most directly relevant to groundborne noise propagation is the resonant column test (RCT). Lo Presti and Pallara[2] published some small strain values for η obtained from resonant column tests which at low frequencies were 0.3 for sand, 0.4–0.8 for clay. The figures for sand were frequency dependent, increasing to 0.4 at 100 Hz. These authors also showed results from cyclic loading torsional shear tests (CLTST), which were about half the values obtained in the RCT tests. Stokoe et al.[3] considered that the reason for RCT damping figures being greater than CLTST figures could be a consequence of the equipment-generated damping in RCT tests, although this only accounts for about 0.1 of the loss factor for clay. It is also possible that the loss factor measured in RCT tests is influenced by viscous losses, and that the loss factor measured in CLTST tests is solely due to friction and structural changes within the soil.

Michaels[4] concluded that shear waves appear to have less damping than p-waves in unsaturated soils. This may be due to the relative reduction in pore fluid density that occurs when water is replaced with air, and this results in less inertia to drive fluid flow through pores. When air is replaced by water, for a given porosity, we should expect a dense fluid to produce greater damping than a less dense fluid, like air. The smaller level of shear wave may be explained by the lack of inertia for a pore fluid composed largely of air.

IMPEDANCE AND BOUNDARIES

An important property of an elastic and a viscoelastic medium is impedance, given here the symbol Z. This is the ratio of force to velocity (its inverse is mobility). It is given by $\sqrt{\rho D} = \rho c$ for p-waves where ρ is the mass density and D is the compression modulus. There is a corresponding expression for shear waves $\sqrt{\rho G} = \rho v$ where G is the shear modulus (also denoted μ). Impedance has the units Ns/m^3 which means that the force is actually pressure in N/m^2. The importance of impedance arises when, instead of an infinite homogenous isotropic space, a wave propagates in a space where two different media are separated by a plane (for example a layer of clay on a rockbed). Assuming for simplicity that each half-space is infinite in depth, and that the wave is purely a p-wave with a flat wave front parallel to the interface plane, when the wave reaches the interface what happens to it depends on the ratio between the impedances of the two media. At the interface, the velocity across the two media must be the same, but the pressure either side of the plane need not be. What happens is that part of the incident wave is reflected at the interface, and part of it is transmitted through it. On the incident side, the incident and reflected wave are superposed (arithmetically added). Since $Z = p/u$, where p is pressure and u is velocity, then the equations governing reflection and transmission are $p_i + p_r = p_t$ and $u_i - u_r = u_t$. The first of these can be written in terms of $Z = p/u$ or $p = uZ$ so that the reflection coefficient is $R = u_r/u_i = p_r/p_i = (Z_t - Z_i)/(Z_t + Z_i)$. The transmission factor T is $1 + R$. An important case is when the medium on the transmission side is a vacuum, that is, Z_i is zero. In this case the reflection coefficient is -1, leading to zero pressure at the surface of the incident plane and velocity doubling. Conversely, if the plane into which the wave is being transmitted is infinitely rigid ($Z_t = \infty$) the reflection coefficient is 1 leading to pressure doubling and zero velocity.

In the corresponding case for shear waves, in which an SV wave is travelling vertically, the same equation applies, after substituting G for D in the calculation of Z.

When a wave is incident on the boundary between an elastic halfspace and a void at an angle of incidence greater than zero, depending on the angle two reflected waves result, one a p=wave and the other an SV wave. The reflection coefficients are given in terms of the ratio κ between the p-wave speed and the SV speed, and the SV wave angle θ_s obtained from Snell's law by $\sin\phi = \sin\theta(v/c)$ where ϕ is the angle for the wave with velocity ν and θ is the angle for the wave with velocity c (Figure 3.1).

The reflection coefficients are given by[5]

$$R_{sp} = \frac{2\kappa \sin 2\phi_1 \cos 2\phi_1}{\sin 2\phi_1 \sin 2\theta_1 + \kappa^2 \cos^2 2\phi_1} \tag{3.15}$$

$$R_{ss} = -\frac{\sin 2\phi_1 \sin 2\theta_1 - \kappa^2 \cos^2 2\phi_1}{\sin 2\phi_1 \sin 2\theta_1 + \kappa^2 \cos^2 2\phi_1} \tag{3.16}$$

$$R_{ps} = \frac{2\kappa \sin 2\theta_1 \cos 2\phi_1}{\sin 2\phi_1 \sin 2\theta_1 + \kappa^2 \cos^2 2\phi_1} \tag{3.17}$$

$$R_{pp} = \frac{\sin 2\phi_1 \sin 2\theta_1 - \kappa^2 \cos^2 2\phi_1}{\sin 2\phi_1 \sin 2\theta_1 + \kappa^2 \cos^2 2\phi_1} \tag{3.18}$$

where R_{pp} is the p-wave reflected from an incident p-wave; R_{ss} is the SV-wave reflected from an incident SV-wave; R_{sp} is the SV-wave reflected from an incident p-wave; R_{ps} is the p-wave reflected from an incident SV-wave and $\kappa = c_p/c_s$ (Figure 3.2).

Another important effect occurs at the boundary between two elastic or viscoelastic halfspaces when the waves concerned are incident at an angle other than 0 degrees. Depending on the angle, not only is there reflection and transmission of both kinds of wave, but there may also be wave conversion in that, for an incident *p*-wave, as well as there being a reflected *p*-wave, there is also a new reflected *s*-wave created, and vice versa. In this case

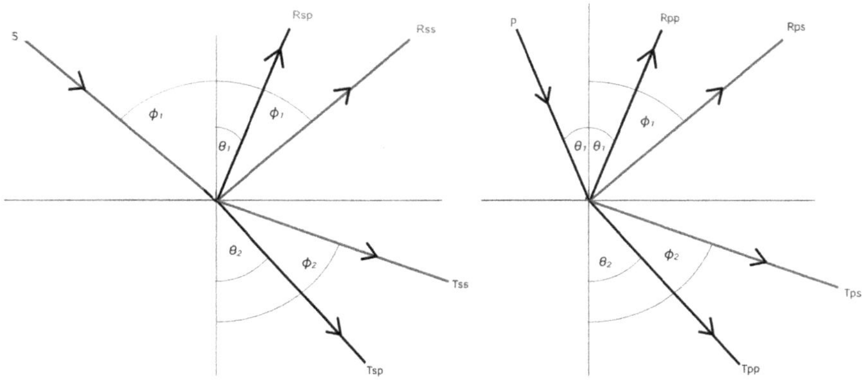

Figure 3.1 Notation for reflected P, SV and transmitted P and SV waves at a boundary between two elastic halfspaces.

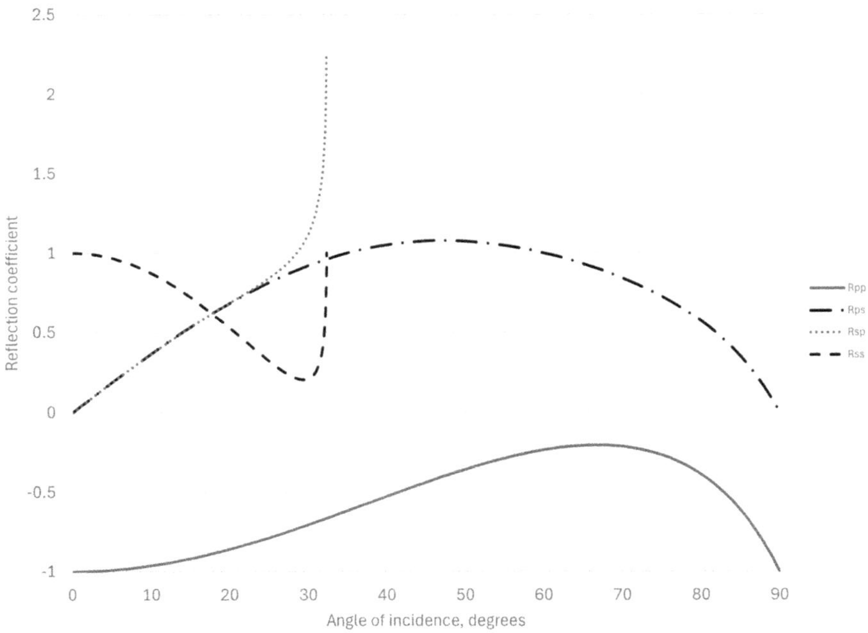

Figure 3.2 Reflection coefficients at the top of a clay halfspace.

refraction occurs, in that the direction of propagation of the transmitted p-wave is not the same as that of the incident wave, nor is it for the incident and transmitted shear waves.

The full set of reflection coefficients, for a perfect bond between two elastic halfspaces (with elastic properties denoted by subscripts 1 and 2) can be calculated as follows.[6]

With the availability of applications capable of straightforward matrix inversion and multiplication (e.g. MDETERM, MINVERSE and MMULT in Excel) the reflection and transmission coefficients are best calculated by way of matrices. The scattering matrix

$$\mathbf{S} = \mathbf{M}^{-1}\mathbf{N}$$

is

$$\begin{bmatrix} R_1pp & R_1sp & T_2pp & T_2sp \\ R_1ps & R_1ss & T_2ps & T_2ss \\ T_1pp & T_1sp & R_2pp & R_2sp \\ T_1ps & T_1ss & R_2ps & R_2ss \end{bmatrix} \tag{3.19}$$

where the subscript 1 indicates the source is in halfspace 1, and 2 is in halfspace 2, and "*ij*" indicates the source is of type i and the reflected or transmitted wave is of type j where i and j are p or s.

Matrix **M** is:

$$\begin{bmatrix} -c_1 p & -\cos\phi_1 & c_2 p & \cos\phi_2 \\ \cos\theta_1 & -v_1 p & \cos\theta_2 & -v_2 p \\ 2\rho_1 v_1^2 p\cos\theta_1 & \rho_1 v_1(1-2v_1^2p^2) & 2\rho_2 v_2^2 p\cos\theta_2 & \rho_2 v_2(1-2v_2^2p^2) \\ -\rho_1 c_1(1-2v_1^2p^2) & 2\rho_1 v_1^2 p\cos\phi_1 & \rho_2 c_2(1-2v_2^2p^2) & -2\rho_2 v_2^2 p\cos\phi_2 \end{bmatrix} \tag{3.20}$$

Matrix **N** is:

$$\begin{bmatrix} c_1 p & \cos\phi_1 & -c_2 p & -\cos\phi_2 \\ \cos\theta_1 & -v_1 p & \cos\theta_2 & -v_2 p \\ 2\rho_1 v_1^2 p\cos\theta_1 & \rho_1 v_1(1-2v_1^2p^2) & 2\rho_2 v_2^2 p\cos\theta_2 & \rho_2 v_2(1-2v_2^2p^2) \\ \rho_1 c_1(1-2v_1^2p^2) & -2\rho_1 v_1^2 p\cos\phi_1 & -\rho_2 c_2(1-2v_2^2p^2) & 2\rho_2 v_2^2 p\cos\phi_2 \end{bmatrix} \tag{3.21}$$

where θ_1 and ϕ_1 are the angles of the P wave and SV wave in halfspace 1 and θ_2 and ϕ_2 are the angles in halfspace 2.

The relationship between θ_1, ϕ_1, θ_2 and ϕ_2 is given by Snell's law from

$$\frac{sin\ \theta_1}{c_1} = \frac{sin\ \phi_1}{v_1} = \frac{sin\ \theta_2}{c_2} = \frac{sin\ \phi_2}{v_2} = p \tag{3.22}$$

where $c_i = [(\lambda_i + 2\mu_i)/\rho_i]^{1/2}$ and $v_i = (\mu_i/\rho_i)^{1/2}$.

In the case of an incident SH wave, the calculation is much more straightforward, as p-waves are not involved at all:

$$R_S = \frac{Z_1\cos\phi_1 - Z_2\cos\phi_2}{Z_1\cos\phi_1 + Z_2\cos\phi_2} \tag{3.23}$$

$$T_s = \frac{2Z_1\cos\phi_1}{Z_1\cos\phi_1 + Z_2\cos\phi_2} \tag{3.24}$$

where $Z_1 = \rho v_1$ and $Z_2 = \rho v_2$.

An important feature of Snell's law is that there is a critical angle when $\sin(\theta)*c_1/c_2 = 1$ and a reflected or refracted wave has an angle of 90°, that is, parallel to the plane separating two elastic halfspaces. At incidence angles at, or greater than, the critical angle, there is no refracted, reflected or transmitted wave.

In the case of transmission of waves from a medium into another with a higher speed of sound, for example from clay into rock, the critical angle affects waves on both sides of the interface. At angles at or greater than the critical angle the refracted wave will travel along the boundary between the two halfspaces as an evanescent wave, that is, attenuated for mathematical rather than energy loss reasons.

Examples of applying Equations 3.20 and 3.21 are given in Figures 3.3 and 3.4.

These results have important implications for transmission of vibration from a source in, for example, clay, above a rockhead. Reflection from the rockhead will occur, but only up to the critical angle, beyond which the reflected and refracted waves are confined to the interface wave at the plane separating the two media.

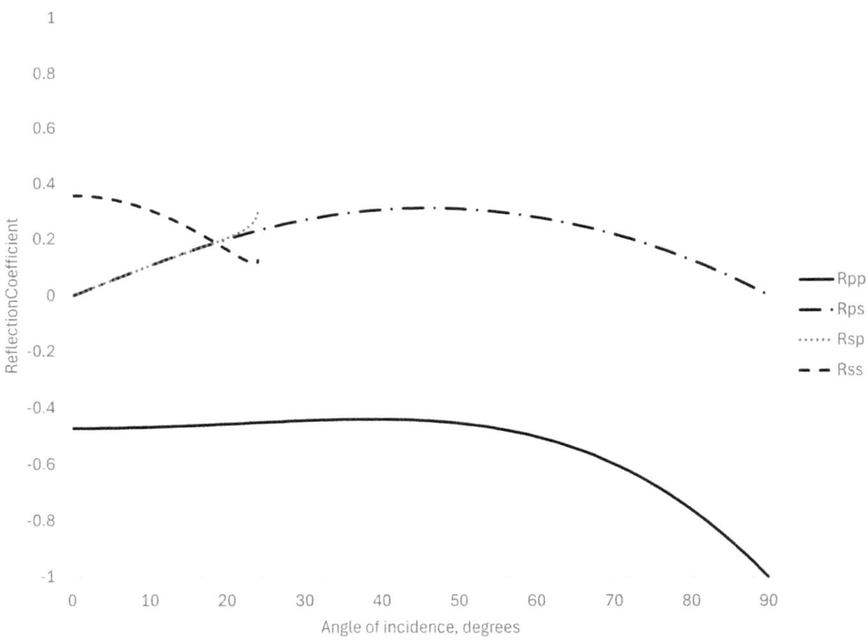

Figure 3.3 Reflection coefficients at the boundary between clay and limestone, source in the limestone.

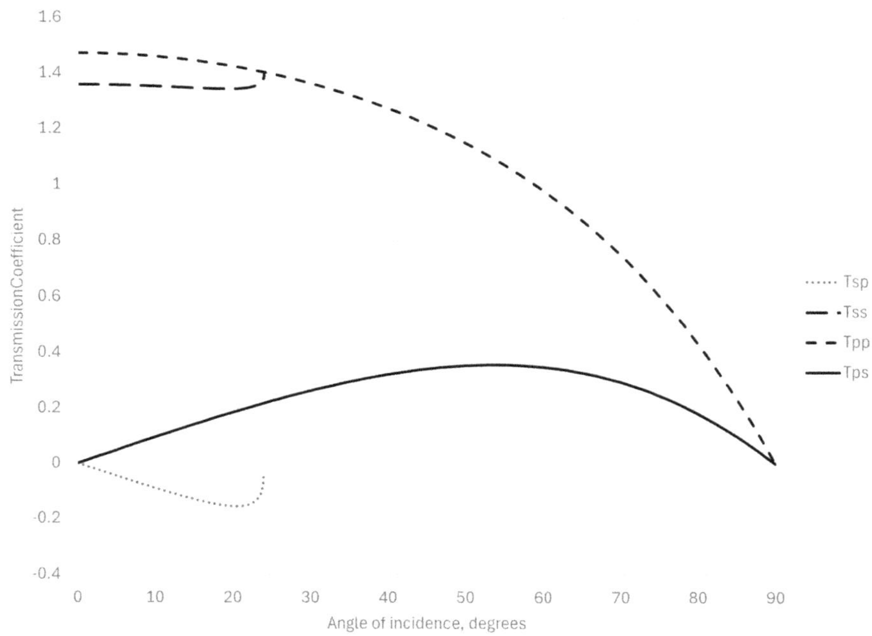

Figure 3.4 Transmission coefficients at the boundary between clay and limestone, source in the limestone.

SPHERICAL AND OTHER NON-PLANE WAVES

The equations for reflection and transmission coefficients apply to plane waves only. While waves originating from a finite source having spherical or cylindrical characteristics may approximate plane waves at a large distance, it is important to take into account that their behaviour at interfaces between layers differs from that of plane waves. An important consequence of this is that, while a plane p-wave incident on the surface of a half-space does not give rise to a Rayleigh (or Scholte) wave at any angle of incidence, this is not true for a spherical p-wave, and a buried spherical source can result in the appearance of a Rayleigh wave on the surface. Equations for reflection and transmission coefficients for incident spherical waves are intractable, and it is necessary to use numerical methods to predict the angle of incidence at which a spherical p-wave results in a Rayleigh wave, and to determine its amplitude.

VISCOELASTIC MEDIA

The equations given above are for an elastic medium. In the case of a viscoelastic medium, that is, where the moduli and wave speeds are complex, the equations are the same, but substituting complex values for c, v and G, which results in a much longer calculation. Many mathematical applications will process complex matrix operations, and there is an Excel add-in available to provide the functions ZMDETERM, ZMINVERSE and ZMMULT to process complex matrices. (It is necessary to use Cramer's rule to avoid the need to process inverse complex matrices). The principal effect is to change the phase of the reflected and transmitted waves, rather than their magnitude.

LAYERED MEDIA

This process of reflection and wave conversion contributes to attenuation of a wave in its progression through an anisotropic medium. In a layered medium it occurs repeatedly.

Algebraically solving the equations for the reflection and transmission coefficients of a system of layered media is a very long process. In geology, clearly distinct boundaries between, for example, soil and rock, are not as common as a progressive increase in modulus and density since rock is frequently overlain by weathered layers, with the amount of weathering increasing with height above the rockhead producing a gradual transition from the properties of rock to the properties of soil. As an example, if it is assumed that there are four layers, (1) competent rock, (2) slightly weathered rock, (3) moderately weathered rock, (4) highly weathered rock and (5) soil, Brekhovskikh[7] gives the following equation for the full set of layers as

$$Z_{in} = \frac{\begin{array}{c} Z_1Z_2Z_3 - Z_1Z_3^2\delta_2\delta_3 - Z_1Z_3Z_4\delta_2\delta_4 - Z_1Z_2Z_4\delta_3\delta_4 \\ -i\left[Z_3^2Z_3\delta_2 - Z_2Z_3^2\delta_3 - Z_2Z_3Z_4\delta_4 - Z_3^2Z_4\delta_2\delta_3\delta_4\right] \end{array}}{\begin{array}{c} Z_2Z_3Z_4 - Z_2^2Z_4\delta_2\delta_3 - Z_2^2Z_3\delta_2\delta_4 - Z_2Z_3^2\delta_3\delta_4 \\ -i\left[Z_1Z_3Z_4\delta_2 - Z_1Z_2Z_4\delta_3 - Z_1Z_2Z_3\delta_4 - Z_1Z_3^2\delta_2\delta_3\delta_4\right] \end{array}} .Z_4 \quad (3.25)$$

In which $\delta_j = \tan\phi_j$ and $\phi_j = k_j \cos\vartheta_j$; k_j is the wavenumber in the j th layer and ϑ_j is the angle of incidence on the j th layer, calculated successively using Snell's law, starting with the angle of incidence on layer 5.

Then the reflection coefficient is $R = \frac{Z_{in} - Z_5}{Z_{in} + Z_5}$.

In the special case of the boundary between an elastic medium and a void, the effect that results when elastic waves meet such a boundary is highly important, albeit mathematically complicated. Surface waves occur, the equations of which were first set out by Lord Rayleigh[8] in 1885, and these are similar to the real-life case at the surface of the ground where the void is filled with air, the equations for which were set out by Scholte[9] in 1947. Rayleigh waves have important characteristics; the first is that they are very shallow and their amplitude diminishes rapidly with depth below ground, and the second is that their propagation speed is always slightly less than the shear wave speed in the medium concerned. Because they do not propagate down into the ground, their geometric expansion is less than is the case for the propagation of an underground p-wave. For a point source, a p-wave has a wave front which expands as the surface of an expanding sphere, and, therefore, its energy is diluted by being spread over an extending surface, and in decibel terms the energy in it diminishes in inverse proportion to the square of the radius (6 dB per doubling). In the case of a point source of Rayleigh waves, the wave front lengthens as the radius of the circle around the point, resulting in an energy dilution in inverse proportion to the radius (3 dB per doubling). If the source of a Rayleigh wave is a line rather than a point, it will not be diminished at all for geometric reasons, whereas an underground point source of p-waves from a line source will be diluted in inverse proportion to the distance (3 dB per doubling). When the propagation medium is lossy, Rayleigh waves, like all waves, will be attenuated according to the loss factor of the medium. When the ground is layered, Rayleigh wave speeds will become frequency dependent.

RAYLEIGH WAVE SPEED

The speed of Rayleigh waves can be found by finding the real root of the equation

$$x^3 - 8x^2 + 24x - 16\gamma x - 16 + 16\gamma = 0 \tag{3.26}$$

in which $x = \rho c^2/\mu$ and $\gamma = \mu/(\lambda + 2\mu) = D$.

This can be done most easily using a goal-seek function such as is found in Excel, to find the value of x, and thereby c, the wave velocity. The value of c is dependent on Poisson's ratio, σ, and is typically about 92–95% of the shear wave speed.

The Rayleigh wave speed can be approximated as

$$c_R = c_S \frac{0.862 + 1.14\,\sigma}{1 + \sigma} \tag{3.27}$$

Rayleigh waves have further features. The particle motion on the surface is elliptical, which is why they are also known as rolling waves (and at the top of the ellipse a particle is actually moving backwards relative to the direction of propagation, earning the description "retrograde"). Consequently, they have oscillating displacements in both the vertical direction and also (to a lesser amplitude) in the direction of propagation. Their amplitude rapidly falls with depth below ground, the depth function being, for a line source

$$U_x \propto e^{-kB_p z} - \left(1 - \frac{c^2{}_R}{2c^2{}_S}\right) e^{-kB_s z} \tag{3.28}$$

$$U_z \propto -B_p \left(e^{-kB_p z} - \frac{e^{-kB_s z}}{1 - \frac{c^2{}_R}{2c^2{}_S}}\right) \tag{3.29}$$

where $k = 2\pi f / c_R$; c_R, c_p and c_S are Rayleigh p-wave and shear wave speeds respectively and $B_p = \sqrt{1 - \frac{c_R{}^2}{c_p{}^2}}$; $B_s = \sqrt{1 - \frac{c_R{}^2}{c_s{}^2}}$.

Equations 3.28 and 3.29 are plotted in Figure 3.5 for the case of a homogeneous clay soil. The depth function is frequency dependent and the depth scale in Figure 3.5 is in units of metres divided by wavelength.

The same data are plotted in Figure 3.6 as a function of depth in metres, for frequencies from 10 Hz to 250 Hz.

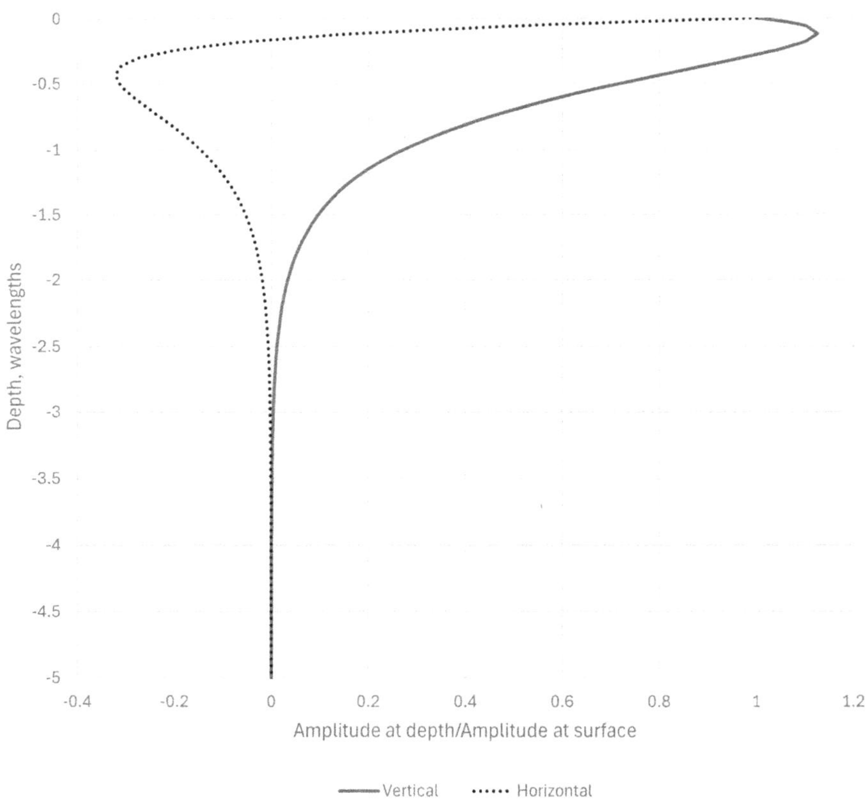

Figure 3.5 Rayleigh wave amplitude as a function of depth in an elastic halfspace.

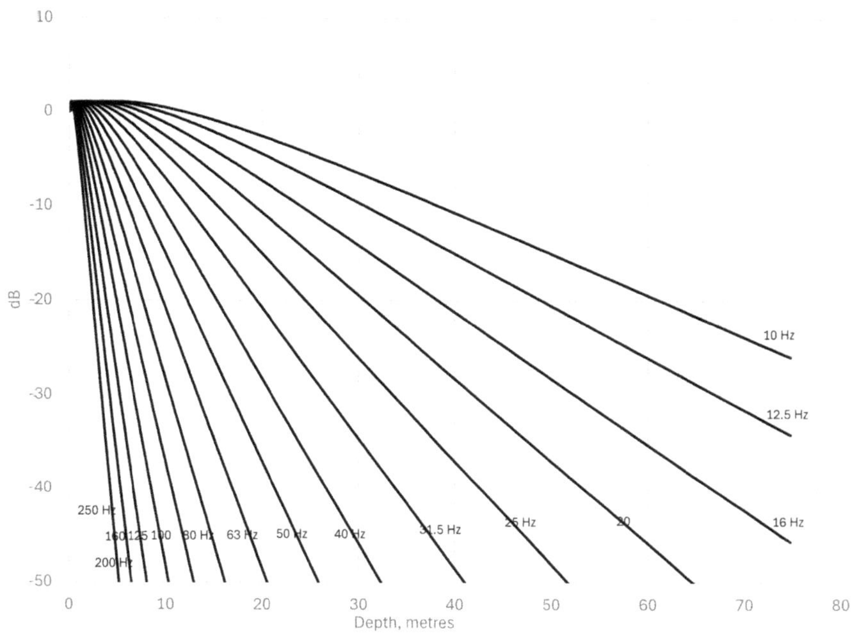

Figure 3.6 As Figure 3.5, in decibels as a function of depth in metres.

While the effect of depth below ground is the same for both line sources and point sources, the effect of distance from the source is different for the two cases. For a line source there is no geometric spreading effect, and, therefore, zero attenuation with distance before taking soil loss factor into account. For a point source the distance function is a Bessel function, the envelope of which gives an amplitude proportional to $1/\sqrt{r}$. Within the envelope there are minima resulting from the form of the Bessel function J (of order 0 for vertical and 1 for radial).

RAYLEIGH WAVES FROM UNDERGROUND SOURCES

While Rayleigh waves may be directly generated by vibration sources on the ground surface, they can also result when p-waves and s-waves from an underground source reach the surface. Pekeris and Lifson[10] found that Rayleigh waves from a buried pulse begin to emerge when the surface distance is greater than 5H where H is the depth of the pulse.

OTHER BOUNDARY WAVES

When the interface is between two elastic halfspaces, rather than between a halfspace and a void, or a halfspace and air, a surface wave of the same kind as a Rayleigh wave occurs that is called a Stoneley[11] wave. Stoneley waves only exist within limits for the ratio of shear moduli and wave speeds.[12] Whether or not a Stoneley wave can exist depends on the values of $\xi = (c_{s1}/c_{s2})^2$ where c_{s1} and c_{s2} are the shear wave velocities in the two half-spaces and $\eta = G_2/G_1$ where G_2 *and* G_1 are the shear moduli. If $\xi \leq 1$ then a Stoneley wave can only exist if $a\eta^2 - b\eta + c > 0$ and if $\xi \geq 1$ it can only exist if $c^*\eta^2 - b^*\eta + a^* > 0$, where

$$a = (2 - \xi)^2 - 4(1 - \xi)^{1/2}(1 - \Lambda_2\xi)^{1/2}$$

$$b = 2\left\{1 - (1 - \xi)^{1/2}\right\}^2 + (1 - \xi)^{1/2}\left[(1 - \Lambda_1\xi)^{1/2}\xi + 4\left\{1 - (1 - \Lambda_2\xi)^{1/2}\right\}\right]$$

$$c = 1 - (1 - \xi)^{1/2}(1 - \Lambda_2\xi)^{1/2}$$

$$a^* = (2 - 1/\xi)^2 - 4(2 - 1/\xi)^{1/2}(1 - \Lambda_1/\xi)^{1/2}$$

$$b^* = 2\left\{1 - (1 - 1/\xi)^{1/2}\right\}^2 + (1 - 1/\xi)^{1/2}\left[(1 - \Lambda_2/\xi)^{1/2}/\xi + 4\left\{1 - (1 - \Lambda_1/\xi)^{1/2}\right\}\right]$$

$$c^* = 1 - (1 - 1/\xi)^{1/2}(1 - \Lambda_1/\xi)^{1/2}$$

$$\Lambda_1 = c_{s1}^2/c_{p1}^2 \quad \Lambda_2 = c_{s2}^2/c_{p2}^2 \qquad (3.30)$$

and c_{p1} and c_{p2} are the p-wave velocities in each halfspace. An asterisk signifies that the replacements $\xi \to \xi^{-1}$, $\Lambda_1 \to \Lambda_2$, $\Lambda_2 \to \Lambda_1$ have been made.

Rayleigh, Stoneley and Scholte waves involve particle motion in the vertical and horizontal directions. There are similar waves that occur less commonly, known as Love waves, which involve lateral particle motion, at right angles to the direction of propagation, and there is also a head wave (also called a von Schmidt wave) which has a depth-dependent horizontal propagation speed, being equal to the p-wave speed at the ground surface, falling to that of the shear wave speed at a depth below ground where its wavefront forms a tangent to the shear wave front.

When there is a plate-like layer below the surface, a bending wave will propagate in the plate known as a Lamb wave. Lamb waves differ from interface waves such as Rayleigh, Stoneley, Love and Scholte waves in that their wave speed is frequency dependent. A wave with a frequency-dependent wave speed is referred to as dispersive (Figure 3.7).

No simple solutions to the wave equations for all but the basic cases explained above are available, and numerical methods are required to carry out analyses of wave propagation in ground conditions involving multi-layered structures.

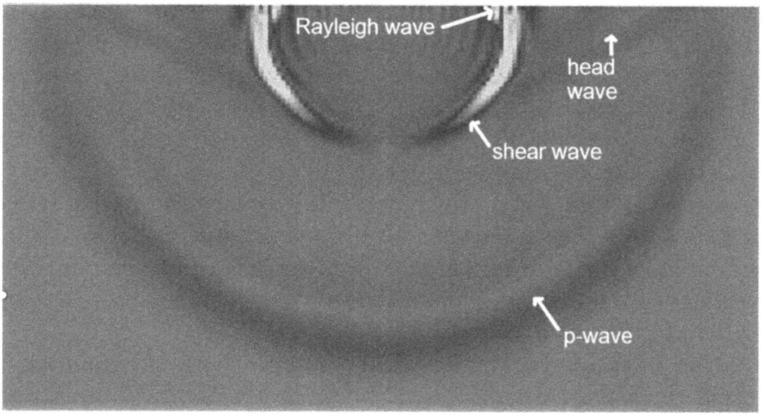

Figure 3.7 Wave types near the surface of an elastic halfspace.

SCATTERING

It is important to take into account the fact that the wave propagation equations discussed apply only to isotropic media and specular (mirror-like) interfaces between halfspaces. In the geological environment, this is never true. There is always some roughness of the surface of a halfspace, or at the interface between two halfspaces, and this has the effect of causing wave scattering. Scattering does not involve the absorption of wave energy, but it does affect reflection and transmission behaviour in a way that influences distance effects. Furthermore, the media in halfspaces are very seldom isotropic, and will contain irregularities which result in scattering.

SURFACE ROUGHNESS

When the surface of a halfspace, on which a Rayleigh wave is propagating, is rough, two effects result.[13] The first is a downward frequency shift and the second is additional distance attenuation.[14]

DISPERSION

The ground is frequently layered to a greater or lesser extent. This may be due to the obvious presence of media of different properties such as layers of sand, gravel or clay, or due to the progressive increase in modulus of a material such as clay with increase in depth below ground.

Layering has two main effects. The first is that it causes depth-dependent variations in the phase velocity of waves, and this results in surface

(Rayleigh) waves becoming dispersive, that is, having phase speeds which are frequency dependent. An important consequence of Rayleigh waves being dispersive is that group waves are formed. The broad principle of group waves is explained in Chapter 7. They consist of a modulated carrier wave that forms wave packets, having a propagation speed different from the non-dispersive Rayleigh wave speed.

The second is that wave propagation downwards through a system of layers differs from the case of a homogeneous isotropic halfspace. Not only is there attenuation due to multiple reflections between layers, but also the ray path of propagation can become curved. This is analogous to a temperature or wind gradient in airborne acoustics.

There are several important practical consequences to the presence in the ground of the wave types discussed, particularly when field measurements are made in order to attempt to discover the magnitude of vibration being emitted by a source, and the dynamic properties of the ground. In addition to the effect that variations with depth in the dynamic properties of the ground have, it is common to find, in soils such as clay, that the dynamic moduli progressively increase with depth. This causes the propagation path of p-waves to be curved upwards towards the ground and this has the effect of enhancing the rate of attenuation due to geometric effects.

Frequently, the ground consists of a soil, such as clay, lying above a layer, such as rock, either limestone sandstone or granite. This complicates the effect of distance from the source when measuring surface waves, because the surface waves as measured will include both the direct wave and the reflected wave when the incident angle is less than critical. When this happens, an interference pattern is created, resulting in areas where, at a particular frequency, the reflected wave is out of phase with the direct wave and partial cancellation occurs when they are superposed.

In very general terms, the geometric effect of distance for a homogeneous halfspace can be summarised as:

- Inverse square law
  Body and shear waves from small sources in unbounded media
- Inverse distance law
  Body and shear waves from line sources in unbounded media
- Inverse $\sqrt{\text{distance}}$ law
  Rayleigh waves from point sources
- Material damping only (no geometric effect):
  Rayleigh waves from line sources

IMPEDANCE OF A SPHERE IN A SOLID MEDIUM

While the impedance of a plane wave in an unbounded isotropic homogeneous medium is a real number, that is, the impedance is a resistance, this

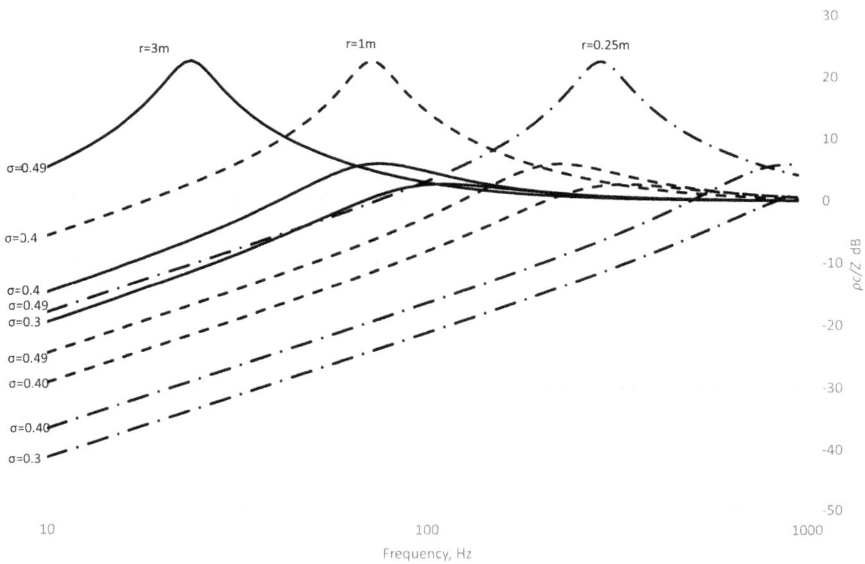

Figure 3.8 Mobility of elastic medium (clay) surrounding a sphere.

changes when the wave is not plane. The simplest case is a sphere, in which case the impedance has a real (resistance) and imaginary (reactance) part. In other words, the force and velocity are out of phase.

The specific radiation resistance of a sphere in an elastic medium is given[15] by

$$Z = \rho c\left(\frac{R^2}{1 + R^2} - i\frac{1 + (1 - K)R^2}{KR\left(1 + R^2\right)}\right) \tag{3.31}$$

where $R = \omega r/c$; r is the radius of the sphere; $c = [(\lambda + 2\mu)/\rho]^{1/2}$ and $K = (1 - \sigma)/2(1 - 2\sigma)$.

When σ is less than one-third, the radiation reactance is a negative or stiffness reactance at all frequencies. When σ becomes greater than one-third, a frequency of zero radiation reactance appears, above which the reactance is positive or inertial. This "resonance" frequency decreases from infinity to zero as σ increases from one-third to one-half.

Some example plots of the mobility $\rho c/Z$ are shown in Figure 3.8.

MOBILITY OF A RECTANGULAR PLATE ON AN ELASTIC HALFSPACE

The exact solution for the mobility of a rectangle on an elastic halfspace of mass density ρ is somewhat intractable, but its vertical response can be

Table 3.1

Aspect ratio	s
1	1.06
1.5	1.07
2	1.09
3	1.13
5	1.22
10	1.41

approximated[16] by assuming the plate has an added mass equal to a body of soil of height $h = 0.27$ times the square root of the base area A. The stiffness K of the soil beneath the rectangle is then given by $E\sqrt{A}\left[s/(1-\sigma^2)\right]$ where E is Young's modulus, σ is Poisson's ratio and s is taken from the following Table 3.1. The viscous damping constant is $5.42\sqrt{\left(K\rho h^3\right)}$.

POROUS MEDIA AND WATER SATURATION

The topic of sound propagation in fluid-saturated saturated porous solids was pioneered by Biot,[17,18] Biot predicted the existence of two compressional waves and one shear wave. One of the compressional waves attenuates very rapidly, and the other can be regarded as a modified form of the compressional wave which would exist in the dry solid.

Biot's work was developed by Stoll[19] to include friction damping in the frame moduli, and was stated in generalised form by Yamamoto and Turget[20] as

$$\left.\begin{matrix} C_{p1} \\ C_{p2} \end{matrix}\right\} = \left[\left(\frac{2(HM-Q^2)}{\left(\rho M + m'H - 2\rho_f Q\right) \mp \left[(m'H-\rho M)^2 + 4\rho_f\left(\rho_f M - m'Q\right)H + 4\rho\left(m'Q - \rho_f M\right)Q\right]^{1/2}}\right)\right]^{1/2} \tag{3.32}$$

where the two complex roots yield values for the fast and slow compressional waves C_{p1} and C_{p2}. The imaginary part of a complex wave speed leads the loss factor through the relationship $\eta = 2 \times \mathrm{Im}[1/C^*]/\,\mathrm{Re}\,[1/C^*]$. where C^* is a complex wave speed.

The Biot-Stoll shear wave velocity is given by

$$C_s = (\mu/\rho)^{1/2}\left(1 - \rho_f^{\,2}/\rho m'\right)^{-1/2} \tag{3.33}$$

H, Q and M are Biot's elastic moduli and related to the basic physical properties of porous media,

$$H = (K_r - K_s)^2/(D_r - K_s) + K_s + 4\mu/3$$
$$Q = K_r(K_r - K_s)/(D_r - K_s)$$
$$M = K_r^{\,2}/(D_r - K_s)$$
$$D_r = K_r\left[1 + b\left(K_r/K_f - 1\right)\right]$$

where K_r is the bulk modulus of the frame material and K_f is the apparent bulk modulus of the pore fluid, and μ and K_s are the complex shear modulus and the complex bulk modulus of the skeletal frame, given by

$$\mu = \mu_r(1 + i\delta)$$

and

$$K_s = K_{sr}(1 + i\delta')$$

in which μ_r is the dynamic shear modulus, K_{sr} is the dynamic bulk modulus, and δ and δ' are the loss factors of the skeletal frame. Also, ρ is the bulk density of the porous medium given by

$$\rho = (1 - \beta)\rho_r + \beta\rho_f$$

where β is the porosity, ρ_r is the density of the frame material and ρ_f is the fluid density.

Finally, m' is the complex virtual mass density given by $m' = (1 + \alpha)\rho_f/\beta - i[\eta F(\omega)/k_s\omega]$, in which α is the added mass coefficient of the frame (usually 0.25), and $F(\omega)$ is a viscous correction factor which, stated in terms of

$$\kappa = \frac{1}{7} d_{mean}\sqrt{\frac{\omega\rho_f}{\nu_f}}$$

as

$$F(\omega) = \frac{\kappa T}{4[1 + 2iT/\kappa]}$$

in which

$$T = \frac{ber'(\kappa) + ibei'(\kappa)}{ber(\kappa) + ibei(\kappa)}$$

where *ber* and *bei* are, respectively, the real and imaginary parts of the Kelvin function of the first kind or order zero, and *ber*′ and *bei*′ are their derivatives. The term ν_f is the fluid dynamic viscosity and d_{mean} is the mean grain size.

The Biot-Stoll formula gives a compressional wave that is always equal to or higher than the conventional elastic compression wave, and a second compressional wave that is slow and highly attenuated. While the second kind of wave can be ignored as a received signal, it is important in that a fraction of the initial wave energy is carried by it, and rapidly attenuated, reducing the amplitude of the fast wave, and also, on reflection at boundaries between soils of different impedances, some of the compression wave energy will be reflected as a wave of the slow kind, also to be rapidly attenuated. This may explain why field measurements of attenuation with distance from ground level or underground sources suggest higher loss factors than can be accounted for by friction, viscous or Masing effects.

By comparison, the Biot-Stoll formula for shear waves is more simply related to the classical shear wave speed for elastic media.

Stoll noted that, when strain amplitudes are less than about 0.0001%, soil moduli can be regarded as linear, and simpler equations for predicting velocity and attenuation are possible. He noted that in the case of dry materials such as sands, silts, sandstones and other granular materials, attenuation in low-amplitude waves may be adequately described by a model which assumes a constant loss factor. The principal mechanism of energy loss in these dry materials is friction that governs the minute amount of slip occurring within the contact area between particles. As a result the damping observed in a granular material is usually very small (e.g. typically of the order of 0.01 in sandstones and sands). When the pore spaces between the particles contain a fluid such as water, the observed attenuation is often significantly larger due to the addition of viscous losses in the fluid. In the case of materials with high permeability, there is overall motion of the fluid field relative to the skeletal frame. This result is frequency-dependent damping which changes rather rapidly over a fairly narrow frequency range. In finer materials with very low permeability, this kind of damping does not appear until quite high frequencies are reached. Another kind of viscous damping occurs, however, even in the low frequency range, as a result of local fluid motion in the vicinity of the intergranular contacts similar in nature to "squeeze film motion", as in the theory of lubrication. This second kind of viscous damping causes frequency-dependent damping that can be described in terms of classical viscoelastic models.

Some examples from using the Biot-Stoll equations for a range of soils and rocks are given in Table 3.2.

Table 3.2 Examples of Biot-Stoll equations for different soils and rocks

	V1	*V2*	*V3*	*Poisson*
Soil (wet)	1955	234	483	0.47
Soil (dry)	971	2.9	519	0.30
Completely weathered granite (wet)	2097	39	351	0.49
Completely weathered granite (dry)	688	8	367	0.3
Highly weathered granite (wet)	2502	73	675	0.46
Highly weathered granite (dry)	1299	9.99	694	0.3
Moderately weathered granite (wet)	5362	139	2833	0.3
Moderately weathered granite (dry)	5396	7.2	2885	0.3
Slightly weathered granite (wet)	4317	142	2265	0.31
Slightly weathered granite (dry)	4277	9.4	2286	0.3
Limestone (dry)	6562	153	3283	0.33
Limestone (wet)	6186	590	3052	0.34
Silty clay (dry)	499	1.3	249	0.33
Silty clay (wet)	1684	5.4	222	0.49
Sand (dry)	5557	215	3399	0.2
Sand (wet)	5518	1255	3352	0.21

An important result from applying the Biot-Stoll equations is that the presence of water saturation in a medium can substantially increase its D modulus and, thereby, increase its Poisson's ratio. For this reason, it is not reliable to compute the value of the shear modulus from either Young's modulus or the compression modulus and Poisson's ratio, if the degree of water saturation in the medium for which these properties were measured is not known.

Some examples of soil properties are given in Table 3.3.

The stiffness of granular soils, represented by the drained Young's modulus, E, or constrained modulus, D, is such that the variation in modulus with depth is proportional to the increase in the N value obtained from the Standard Penetration Test or SPT. N is the number of blows required for a sampler to penetrate 300 mm of soil. When normalised to 60% hammer efficiency, the result is given by the symbol N_{60}. Soil modulus increases with depth, and there is a relationship between E and uncorrected N_{60} values.

Many authors have sought to use the SPT N value for estimating the Young's modulus, N_{60}, of granular soils. Although a number of different equations have been used, most of these correlations have the general form of

$$E = a(N + b)$$

where E = Young's modulus; N = SPT blow counts; a and b = empirically derived constants.

Table 3.3 Examples of soil properties

	Shear modulus, G *Gpa*	*Compression modulus, D* *Gpa*	*Density* kg/m^3	*Loss factor*
Alluvium (Silt)	0.07	0.27	1800	0.02
Chalk	2.00	14.60	2000	0.03
Chalk Layer 1	0.72	3.68	1750	0.05
Chalk Layer 2	0.74	3.78	1800	0.05
Chalk Layer 3	0.86	4.42	2100	0.05
Chalk Layer 4	2.00	14.58	2000	0.05
Chalk Layer 5	7.50	28.32	2450	0.05
Concrete	20.80	55.60	2400	0.02
Gault Clay	0.21	3.67	1900	0.05
Gravel	0.27	1.03	2000	0.05
Gravels (terrace)	0.27	1.04	2000	0.05
Hard Chalk	2.00	14.60	2000	0.03
Harwich Formation	0.16	2.00	1930	0.05
Highly weathered	0.19	1.14	2320	0.05
Limestone	2.26	13.69	2700	0.05
London Clay	0.74	4.41	1700	0.05
Made Ground	0.02	0.19	1500	0.09
Fresh Granite	22.66	89.48	2660	0.05
Granite (slightly weathered	21.30	79.70	2540	0.05
Granite (moderately weathered)	12.20	56.00	2430	0.05
Mudstone (unweathered)	0.75	2.00	2000	0.05
Mudstone (weathered)	0.03	0.13	1840	0.06
Peat	0.05	0.20	1000	0.03
Ragstone	2.26	13.69	2300	0.05
Residual Soil	0.50	3.01	2090	0.05
Rock	13.16	49.20	2600	0.01
Sand	0.45	5.79	1930	0.1
Sand (deep)	0.19	0.51	1700	0.02
Sand (shallow)	0.09	0.23	1700	0.02
Stone	1.50	4.50	2500	0.05
Terrace Gravels	0.27	1.03	1800	0.06
Thanet Sand	0.45	5.79	1930	0.07
Water	0.00	2.37	1020	0.001

A range of authors' reported correlations between N and E are presented in Table 3.4.

It is apparent from the material presented in this chapter that the characteristics of propagation of vibration in the ground, except in simple cases, are difficult to calculate. Simple approximations are likely to involve uncertainty of considerable magnitude. In cases where there is a complex system of

Table 3.4 Reported equations for deriving soil modulus from N

Equation	*Soil type*	*References*
E (MPa) $= 7(N)^{05}$	Sand	Denver
E (MPa) $= 3.5N$ to $40N$	Sand	Clayton
E (MPa) $= 2.22N^{0888}$	Partially saturated gravels	Wrench & Nowatzki
E (kPa) $= 1200(N + 6)$	Gravelly sand and gravel	Bowles
E (kPa) $= 320(N + 15)$	Clayey sand	Bowles
E (kPa) $= 300(N + 6)$	Silty sand	Bowles
E (MPa) $= 1.6N$	Residual	Jones & Rust
E (MPa) $= 7.5 + 0.8N$	Sand	Papadopoulos
E (MPA) $= 2.5N_{60}$	Saprolite	Decourt

viscoelastic media, including layers and boundaries between different media, it is usually necessary to employ numerical methods to simulate the behaviour of such systems, and numerical modelling is the subject of Chapter 10.

GREENS FUNCTION

The dynamic behaviour of complex systems such as layered soils and rocks and building structures can be difficult to simulate due to lack of detailed and accurate information about the dynamic properties of the material involved. One solution to lack of information is in-situ measurement. In particular, measurements can yield information about soil loss factors, and the behaviour of layers, and topics such as coupling loss factors between the ground and buildings.

These parameters are frequency dependent, and it is, therefore, necessary to obtain measurements across the range of frequencies of interest. This raises a key question as to what the vibration source should be. There are two considerations when it comes to deciding what the source should be. The first is the matter of frequency and the second it its geometrical properties. A measurement of the propagation of vibration from a test source to a receiver will depend on the size and shape of the source, since the driving point impedance of an element is not the same as the impedance of an area source.

Dealing firstly with the question of frequency, there are two options. One is to use an oscillating source such as an electomagnetic exciter, and the other is to use an impulse source such as a hammer or dropped weight, or an explosive charge. The great advantage of an impulse source is that it contains frequencies across a wide band. The theoretical extreme of an impulse, namely a Dirac function, is an infinitely large impulse of infinitely short duration such that its time integral is 1. Importantly, its frequency spectrum is also 1 at all frequencies. In the real world, such impulses do not exist, but even a finite impulse of finite duration has a very broad and flat frequency spectrum. Thus, the transfer function between an impulsive source and a receiver can provide information about the vibration transmission properties of the system over a very broad frequency range.

Likewise, when carrying out numerical simulations in the time domain, an impulse source can be used to similar effect. It has to be borne in mind that, while a true Dirac function is zero for all time except $t = 0$ when it is infinite, in a numerical simulation the frequency transform will be carried out using Fourier transform, which assumes that the events within the transform window are repeated indefinitely with a period equal to the time duration of the window. This effect is highlighted by comparing the spectrum obtained by a Laplace transform, which assumes that the signal does not repeat itself, and a Fourier transform which is periodic in nature.

The impulse response of a system is important not only in acoustics and vibration, but also to electronic circuits. It is also known as a Green's function.

NOTES

1. Boltzmann, L., "Zur Theorie der elastischen Nachwirkung", *Annalen der Physik*, 7, 624–654, 1876.
2. Lo Presti, D.C.F., and Pallara, O. "Damping ratio of soils from laboratory and in situ tests." In *Seismic behaviour of ground and geotechnical structures*, Sèco e Pinlo, Q, (ed.), Balkema, Rotterdam, 1997
3. Stokoe, K.H.I., Hwang, S.K., l-ee, J.N.-K., and Andrus, R. "Effects of various parameters on the stiffness and damping of soils at small to medium strains." *Keynote Lecture 2*, IS Hokkaido 1994 2, 785–816, 1995.
4. Michaels, P. "Comparison of viscous damping in unsaturated soils, compression and shear." *Unsaturated Soils 2006* (GSP 147), ASCE Publications, Reston, VA, 2006.
5. Auld, B.A., *Acoustic fields and waves in solids*, Volume II, 2nd Ed., Krieger Publishing Company, Malabar, FL, 1990.
6. Knott, C.G. III. "Reflexion and refraction of elastic waves, with seismological applications." *The London, Edinburgh, and Dublin Philosophical Magazine and Journal of Science*, 48(290), 64–97, 1899, https://doi.org/10.1080/14786449908621305
7. Brekhovskikh, L.M. *Waves in layered media, applied mathematics and mechanics*, Volume Six, Academic Press, New York, 1960.
8. Rayleigh, L. "On wave propagated along the plane surface of an elastic solid." *Proceedings of the London Mathematical Society*, s1-17(1), 4–11, November 1885.
9. Scholte, J.G. "On Rayleigh waves in visco-elastic media." *Physica*, 13, 245–250, 1947.

10. Pekeris, C. J., and Lifson, H., "Motion of the Surface of a Uniform Half-Space Produced by a Buried Pulse, *The Journal of the Acoustical Society of America* 29(11), 1233–1238, November 1957.
11. Stoneley, R., "Elastic waves at the surface of separation of two solids." *Proceedings of the Royal Society*, 01 October 1924, https://doi.org/10.1098/rspa.1924.0079
12. Chadwick, P., and Borejko, P. "Existence and uniqueness of Stoneley waves." *Geophysical Journal International*, 118, 279–284, 1994.
13. Eguiluz, A.G., and Maradudin, A.A. "Frequency shift and attenuation length of a Rayleigh wave due to surface roughness." *Physical Review B*, 28(2), 728, July 1983, Condensed Matter.
14. Sarris, G., Haslinger, S.G., Lowe, M.J.S., Huthwaite, P., and Nagy, P.B. "Attenuation of Rayleigh waves due to three-dimensional surface roughness: A comprehensive numerical evaluation." *The Journal of the Acoustical Society of America*, 154(2), 808–818, August 2023.
15. Blake, F.G. Jr., "Spherical wave propagation in Solid Media." *The Journal of the Acoustical Society of America*, 24, 211–215, 1952, https://doi.org/10.1121/1.1906882
16. Newmark, M., and Rosenblueth, E. *Fundamentals of earthquake engineering*, Prentice-Hall, Englewood Cliffs, NJ, 98, 1971.
17. Biot, M.A. "Theory of propagation of elastic waves in a fluid-saturated porous solid – I low frequency range frequency rage." *The Journal of the Acoustical Society of America*, 28(2), 168–191, 1956.
18. Biot, M.A. "Theory of propagation of elastic waves in a fluid-saturated porous solid – II higher frequency range." *The Journal of the Acoustical Society of America*, 28(2), 179–191, 1956.
19. Stoll, R.D. "Marine sediment acoustics." *The Journal of the Acoustical Society of America*, 77(5), 1789–1799, May 1985.
20. Yamamoto, T., and Turgut, A., "Acoustic wave propagation through porous media with arbitrary pore size distributions." *The Journal of the Acoustical Society of America*, 83(5), 1744–1751, May 1988.

Chapter 4

Receivers of groundborne noise

Although it can in some cases be heard out-of-doors, groundborne noise is nearly always considered when it is heard in buildings. Frequently, it is necessary to predict how it will be heard in a building that does not yet exist. In those cases, the source may already exist and vibration in the ground at or near where the proposed building will be constructed can be known. In some cases, neither the source nor the building exists at the time when predictions are required.

When the source, but not the building, exists, vibration in the ground can be measured, but that signal will be modified significantly when the building is constructed, both as a result of the effect of the presence of the building on the vibration at the point of entry, and as a result of the transmission chain from soil to foundations, from foundations to structure and finally from structure to air, where it is manifested as airborne sound.

The first question to be answered is what kind of wave is producing the without-building vibration in the ground? If the source is on the surface and the ground is homogeneous and isotropic to a great depth the wave will be almost entirely a Rayleigh wave. As explained in Chapter 3, the amplitude of Rayleigh waves is depth dependent.

Figure 4.1 shows the same data as that in Figure 3.6 of Chapter 3 but expresses the amplitude ratio in decibels.

If the ground is not homogeneous, and there is either a rockhead some metres below ground level, or soil moduli that increase progressively with depth, the wave type at the surface will contain a contribution from p-waves. It should be noted that it was shown in Chapter 3 that the water content of soil affects the compressive modulus to a much greater extent than the shear modulus. If the vibration at the surface contains a significant contribution from p-waves the effect of depth will not be the same as the case for pure Rayleigh waves. It will be dependent on the angle of incidence of the p-waves, and, as it was explained in Chapter 3, at grazing incidence the reflected p-wave tends to cancel the direct wave, whereas at normal incidence, the reverse is true. At each reflection of the p-wave either from the rockhead or from the surface, as well as a reflected p-wave, a new reflected s-wave will be generated.

DOI: 10.1201/9781003618775-4

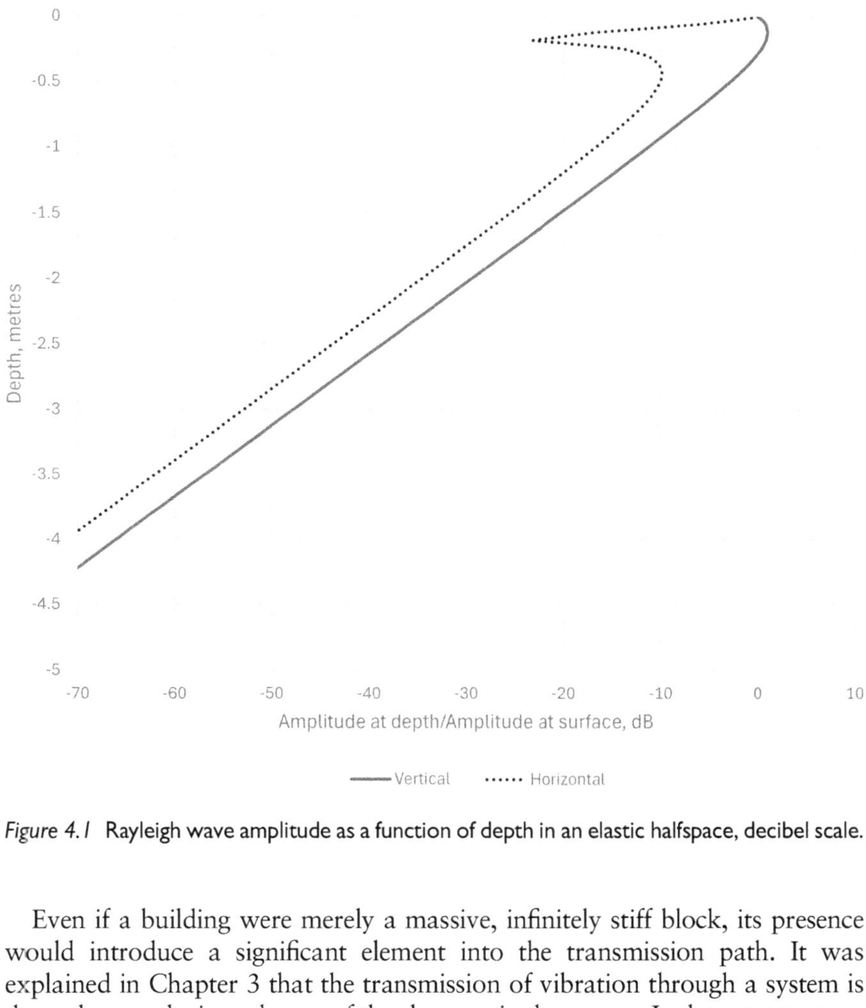

Figure 4.1 Rayleigh wave amplitude as a function of depth in an elastic halfspace, decibel scale.

Even if a building were merely a massive, infinitely stiff block, its presence would introduce a significant element into the transmission path. It was explained in Chapter 3 that the transmission of vibration through a system is dependent on the impedances of the elements in the system. In the extreme case, the terminating impedance can influence the input impedance of the system as seen by the radiating source to the extent that it affects the power transmitted out of the source into the system. At the very least, the impedance of the receiving element, in this hypothetical case being a massive, infinitely stiff lump, will determine how the receiving element responds to incoming vibration.

It was explained in Chapter 3 that a rectangle on an elastic halfspace of mass density ρ has a vertical response that can be approximated by assuming the plate has an added mass equal to a body of soil of height $h = 0.27$ times the square root of the base area A. The stiffness K of the soil beneath the rectangle is then given by $E\sqrt{A}\left[s/\left(1-\sigma^2\right)\right]$ where E is Young's modulus, σ is Poisson's ratio and s which varies between 1.06 and 1.42 according to the aspect ratio of the sides of the rectangle in the range 1 to 10. The viscous damping constant is $5.42\sqrt{\left(\mathrm{K\rho h^3}\right)}$.

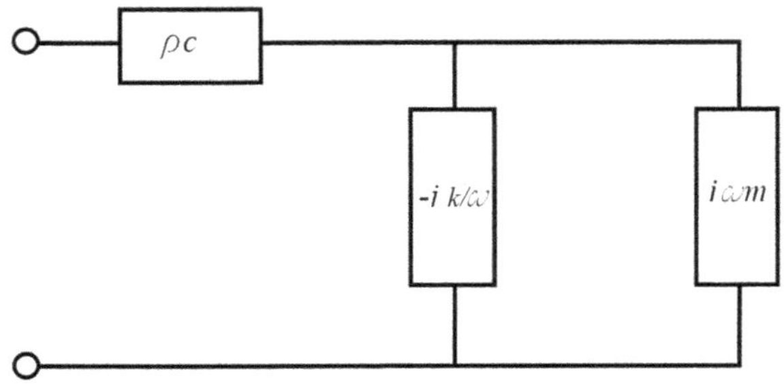

Figure 4.2 Circuit representation of transmission from soil to a mass on a spring.

Using this information, the starting point is to look at the behaviour of the Single-Degree-of-Freedom, mass-on-a-spring case, which may be set out as a circuit as shown in Figure 4.2.

The ratio of the current through the circuit with impedance $i\omega m$ mounted on a spring with impedance $-ik/\omega$ to the current through a resistance of ρc is equal to $\frac{1}{1-\frac{i}{\rho c\left(\frac{\omega}{k}-\frac{1}{\omega m}\right)}}$. At resonance, were it not for damping, the introduction of the building would result in the current through the soil ρc dropping to zero. The current, that is, the velocity, through the building $i\omega m$, for a 10 m × 30 m five-storey concrete building is such that the coupling loss factor, defined as the ratio of the vertical velocity of the building to the open site vertical velocity of the ground, is large, approximately 24 dB at 10 Hz, increasing at the rate of 6 dB per octave.

In reality, the impedance of the building will not simply be $i\omega m$ because it will have many resonant modes due to the presence of many spring-like elements in the structure. The load on the columns of the building is high enough for them to behave like springs, even though they are concrete, and ignoring the modes of the floor plates, this creates a ladder network of masses and springs.

Since criteria for groundborne noise and vibration apply to locations where building occupants will perceive them, it is necessary to convert vibration acceleration measurements obtained on the surface of the undeveloped ground into sound levels inside a future proposed building in dBA. There are four stages in this process. The first is the soil-structure interaction/coupling loss factor that takes account of the effect of the presence of the future building on the ground and the efficiency with which external vibration is transferred into the building. The second stage is the effect of resonances in the structure, particularly floor plates, which have an amplification effect. The third stage is the floor-to-floor attenuation with increase in floor level in the building. The fourth stage is the conversion of vibration in the building surfaces into sound in the air in the rooms enclosed by the surfaces. None of these three stages is capable of exact algebraic solution as the equations involved are nearly all intractable. The available alternative approaches are either to take figures from measured coupling loss factors and so forth, in similar buildings and apply them to the proposed building, or to carry out numerical modelling by the creation of a finite-element or finite-difference model of the structure and simulation of the dynamic behaviour of the system. The former approach has been formalised in guidance documents published by the US Federal Transit Administration (FTA). Further information can also be found in reports prepared as part of the RIVAS and SILVARSTAR[1] programmes. The FTA issued guidance documents in 2006 and 2018 FTA that are not identical.

COUPLING LOSS FACTOR AND FLOOR AMPLIFICATION

One of the most widely cited publications which addressed the topic of coupling loss factor is the "Handbook of Urban Rail Noise and Control",[2] on which the FTA guidance is based. It presents curves for coupling loss factor for types of building design that have been reproduced in many later documents, including the FTA publication "Transit Noise and Vibration Impact Assessment"[3] and the *Transportation Noise Reference Book*.[4] It introduces the subject with the explanation:

> A given building's response to ground vibration is very complex and perhaps impractical to predict analytically. The building foundation is necessarily distributed over a relatively large area in comparison with the wavelengths of vibration in soil for audible frequencies. It may therefore not be valid to model the building as a rigid body on an elastic half-space. Further complicating the problem, the vibration of the ground cannot be viewed as independent of the building foundation vibration.

Four basic types of building foundations are considered: (1) slab-on-grade, (2) spread-footings, (3) piles founded on earth, and (4) piles supported on rock. In the case of slab-on-grade floors the coupling loss is given as 0 dB at low frequencies up to the resonance frequency of the slab on the soil. For the other cases, curves of building foundation vibration relative to ground surface vibration are presented, ranging from −5 dB to −15 dB. The question of whether the ground surface vibration is that with or without the building present is answered in a subsequent paper[5] "A Prediction Procedure for Rail Transportation Groundborne Noise and Vibration". Citing the original curves from the *Handbook of Urban Rail Noise and Control*, the authors explain that

> the foundation response is the level of actual foundation vibration relative to the level of incident ground surface vertical vibration that would exist in the absence of the building structure and its foundation. The appropriate response is added to the estimated ground surface vibration levels to estimate building foundation vibration. No correction is applied to the ground surface vibration to estimate basement floor and wall vibration or vibration of slab-on-grade floors.

Building coupling loss factors, floor-to-floor attenuations and floor amplification factors can only be determined numerically with a reasonable margin of uncertainty. However, it is helpful to review the basic principles involved. There are normally two cases to consider: (1) sources on or near the ground surface which primarily give rise to Rayleigh waves, and (2) sources below ground level which mainly emit p-waves and SV-waves. With regard to case 1 it is rare that the ground below the surface is a semi-infinite homogeneous isotropic half-space, and that means that there will often be, in addition to Rayleigh waves, p-and s-waves resulting from reflections from layers in the ground below the surface.

In Chapter 3, the equations governing the propagation of Rayleigh waves were set out and the very important effect of depth below ground presented. Also in Chapter 3 the reflection of p-waves and s-waves form an underground source was discussed, and an important aspect in that case is that if the angle of incidence of a wave from an underground source is close to zero (i.e., normal incidence), the result of reflection from the surface is to cause a doubling in the amplitude of the particle velocity (and therefore acceleration and displacement). This is caused by the fact that the impedance of air, being the other half-space above the ground, is so low in comparison to that of the ground that the reflection coefficient is close to −1. When a building is placed on top of the halfspace, the impedance of the region above the surface of the halfspace obviously changes significantly. The manner in which it changes depends on the type of structure involved.

Building foundations where the lowest floor level is close to ground level range from raft foundations, that is, a concrete slab lying on the ground, a system of groundbeams supporting concrete slabs, concrete pads above which slabs are cast, and piles with pile caps supporting concrete slabs. Above these foundations may be a framework of columns supporting floor plates. The columns may be concrete having a range of cross sections from square to "blade", or steel "I" sections, and the structure they support may be reinforced concrete slabs with or without beams, or a steel frame of a range of sections. Concrete slabs may be cast over a filler layer of insulating/void-creating compressible material. Other methods of construction include strip foundations supporting walls of brickwork between which span timber floors on joists.

Fundamentally, there are three kinds of mechanism responsible for coupling loss factor. The first is the effect of changing the impedance of the airspace above an elastic halfspace from near zero to a finite value by inserting an element into the airspace. The simple case is if the impedance of the inserted element has an impedance equal to that of the halfspace, no reflection will occur and p-waves and s-waves will be transmitted into the element without change. This results in a coupling loss factor of −6 dB at all frequencies due to the removal of the velocity-doubling effect. If the inserted element has an impedance higher than that of the halfspace, the reflection coefficient changes sign, and in the extreme case will approach +1. In there will be near pressure doubling rather than velocity doubling, with velocity tending to zero and the coupling loss factor changing from −6 dB to a value tending to infinity. In a building, high impedance can be a result not only of high mass, but of resonance in modes within the building structure.

The second kind of mechanism is, therefore, the presence of modes in the structure. At the natural frequency of a mode, the coupling loss factor will dip down by many dB, but the value at the lowest point in the curve will be governed by the loss factor in the system constituting the building. It is also affected by the mass of the building, but that has a greater effect on the frequency of the maximum loss rather than the value of the loss. Although the influence of loss factor is greater with increasing frequency, the modal density of the structure will increase with frequency, so that the part of the coupling loss factor attributable to this effect diminishes in magnitude towards the higher frequency end of the spectrum.

The third kind of mechanism results from the relationship between the wavelength of the incident wave and the size and mode shapes of the building foundation. If the incident wave has a wavelength greater than twice the building width or length, in terms of vertical excitation, the exciting forces acting on the foundation will all have the same sign. At shorter wavelengths/high frequencies there will be parts of the foundation that are being excited with a force of the opposite sign, depending on the angle of incidence. If the building foundation were infinitely rigid and the incident forces were summed across its length and breadth, there would be a cancelling effect, resulting in a fall in the net amplitude for the entire foundation with increasing frequency. In reality the foundation is not rigid and exhibits a modal response leading to a more complex effect.

In the case of a concrete slab laid directly on to the soil (which is actually not a common method of construction), several authors[6,7] have addressed the coupling that occurs in this case. For an infinite slab, the coupling loss is near zero up to the coincidence frequency, being the frequency at which the bending wavelength of the slab coincides with the horizontal wavelength (i.e., the wavelength along the ground surface) of the input wave. At higher frequencies there is a steep attenuation curve, with attenuation increasing with frequency.

Slabs are not, of course, infinite, and when finite slabs are considered, not only does the effect of a finite region of waves in the ground containing a combination of incident and reflected p- and s-waves have to be taken into account, but also the slab will have a set of resonant modes, albeit damped by the contact with the soil beneath. Numerical solution is the only available approach for these cases.

The need to know the coupling loss factor actually only arises in limited cases. Given that it is a correction to apply to ground vibration at the location of the building of interest, without the presence of the building, in order to arrive at the amplitude of vibration in the building foundations, the question arises as to when the vibration of the building foundations is of interest. Cases where it might be necessary to know building foundation vibration are limited to those where foundation vibration has been specified, or where access to the interior of the building may be unavailable, but in nearly all cases the matter of interest is vibration of surfaces within the building, and the re-radiated airborne sound that they produce. To predict the amplitudes of vibration of interior surfaces, such as floors, it is necessary to apply not only the coupling loss factor, but also the amplification of vibration that occurs as a result of resonant modes in the surfaces concerned.

At this point it is helpful to bear in mind that the main influence of the coupling loss factor of a building is modes of vibration within the building structure. If the building were a solid, rigid mass, the coupling loss would be more readily determined. It is the presence within the building of modes of vibration that causes peaks and troughs in the spectrum of its impedance. Floor plates in a building may be regarded as a network of mass-spring systems. For each mode, the floor can be assigned a "modal mass" (only readily capable of calculation in the case of a simply-supported plate), and the bending stiffness of the plate will provide the spring to create a mass-spring system.

When a mass-spring system is in resonance, for a constant-amplitude sinusoidal exciting force, it will build up to an infinite amplitude if zero damping is assumed. This will give it an infinite impedance, and the coupling loss factor at the foundation supporting the spring will be infinitely large. The presence of damping makes the amplitude and the impedance finite. In a simple building structure with a floor plate supported by columns, exciting its foundation so as to set up a resonant response in the plate at the frequency of a plate mode will result in high amplitude vibration of the mid-span of the plate, but low amplitude vibration in the foundation. The "coupling loss" will be very significant. However, while the amplitude of the foundation with the building in place relative to the amplitude of the of the same foundation without the building present may be very low, the amplitude of the mid-span vibration of the plate will be a very large number multiplied by a very small number, resulting in either unity, or even a number greater than one, if energy from a continuous constant-amplitude causes energy storage in the system. In this kind of simple case, there is clear justification for combining the traditionally used two corrections of coupling loss factor and floor amplification into one correction and making the correction zero in decibel terms. In a more complex structure, there will be very many modes, some of which involve structural responses that are not modes in floor plates, or surfaces radiating groundborne noise. This may result in a combined coupling loss/amplification factor which is a net reduction.

An important consideration is that the effect of floor amplification is at its maximum at the location of antinodes associated with the relevant mode. Measurements are often made at only a few locations, and it would be rare to obtain a result for the position of the antinode for each mode within the audible range. In the case of simply supported plates, the positions of the antinodes are readily determined, but with more complex support such as clamped edges, and with point support at column positions across the plate, the positions of antinodes can only be estimated and, in most practical cases, measurements do not capture the maximum floor plate response. By contrast, all the antinodes do contribute to airborne sound radiated by the vibration plate, although a similar problem arises with measurement of airborne noise in that, for the room sizes usually affected by groundborne noise, the sound field will also be modal. Although the mode shapes of room modes in the air within a room are quite different from those of vibration in the plates (floor, walls and ceiling) that are generating the airborne sound, they similarly involve nodes and antinodes, and measured sound levels in a room containing groundborne noise will vary widely with position and height in the room. This is discussed in more detail in Chapter 8.

It is possible to assign to a plate a mobility based on the characteristic mobility of an infinite plate as defined in EN 15657[8]

$$Y_{char} = \frac{1}{8\sqrt{B'm'}} \tag{4.1}$$

where m' is the mass per unit area and $B' = \frac{Eh^3}{12(1-\mu^2)}$ is the bending stiffness for modulus of elasticity E, Poisson's ratio μ and plate thickness h.

Figure 4.3 shows[9] measured point mobility Y at seven locations on a dense concrete plate. The modal peaks and troughs in point mobility are difficult to predict, but it is possible to estimate the limit[10] $Y_{\max}$ and $Y_{\min}$ from

$$Y_{max} = Y_{char}\coth\left(\frac{1}{\beta}\right); Y_{min} = Y_{char}/\coth\left(\frac{1}{\beta}\right); \tag{4.2}$$

where

$$\beta = 4/(\omega\eta m Y_{char})$$

Dependent on the loss factor η, angular frequency ω and mass m.

$Y_{\max}$ and $Y_{\min}$ are plotted in Figure 4.3.

For a plate excited near borders or a corner, the mobility is as follows, first for a rigid edge at distance a:

$$Re\{Y_i\} = Re\{Y_{i,\infty}\}[1 - J_o(2k_B a)] \approx Re\{Y_{i,\infty}\}$$
$$\left[(k_B a)^2 - \frac{1}{4}(k_B a)^4\right] \text{ for } k_B a \leq 1 \tag{4.3}$$

And second for a rigid corner at distance a, b:

$$Re\{Y_i\} = Re\{Y_{i,\infty}\}\left[1 - J_o(2k_B a) - J_o(2k_B b) + J_o\left(2k_B\sqrt{a^2 + b^2}\right)\right]$$
$$\approx Re\{Y_{i,\infty}\}\frac{(k_B a)^4}{2}\left(\frac{b}{a}\right)^2 \text{ for } k_B a \leq 1 \tag{4.4}$$

In the case of lightweight building elements the mobility at the fixing points on frames or ribs can be approximated by[11]

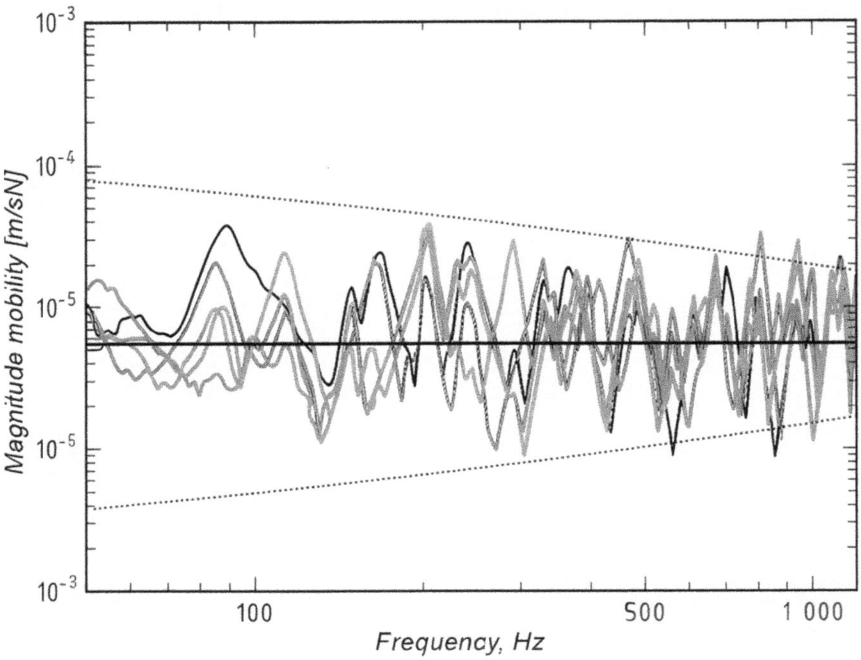

Figure 4.3 Point mobility at seven locations on concrete plate.

$$Y_{beam} = \frac{1}{2m' c_B (1 + j)} \tag{4.5}$$

where c_B is the bending wave velocity and m' is the mass per unit length.

Figure 4.4 shows[9] the measured point mobility over a timber-joist floor. Beam-like behaviour is evident at low frequencies and plate-like behaviour at high frequencies, with a monotonic increase in the transition region.

A number of authors have presented ranges of measured data.

Lurcock and Thompson[12] developed a set of empirical formulae based on the results of a finite element model study. $L_{v,col}$ is estimated from:

$$L_{v,col}(f,n) = L_{v,base}(f) + n/N\, C_1(f) + nC_2(f) \tag{4.6}$$

where $L_{v,base}(f)$ is the frequency dependent vibration level at basement level; n is the storey index of the floor of interest, from basement ($n = 0$) to roof level ($n = N$); N is the total number of storeys in a building, including any basement levels; $C_1(f)$ is a correction factor for the fundamental vertical building mode; and $C_2(f)$ is a correction factor to account for damping loss at each storey.

$C_1(f)$ is defined as

$$C_1(f) = C_{1,1} e^{-\left(\frac{f - f_0}{C_{1,2}}\right)^2} \tag{4.7}$$

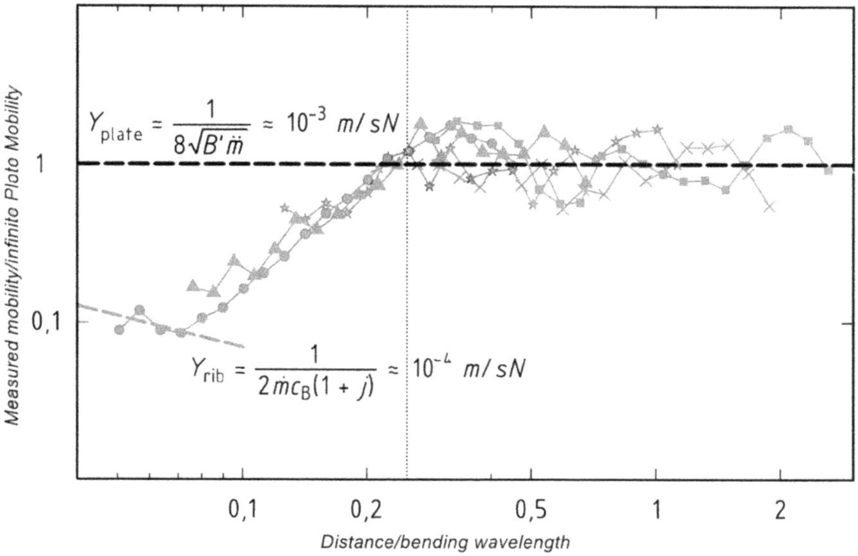

Figure 4.4 Real part of normalised point mobility, as function of normalised distance to fixing.

where f_0 is the fundamental vertical mode frequency, in Hz. $C_{1,1}$ is an amplitude factor, which will represent the total amplification in decibels at the roof level (relative to basement), *at* f_0. For the generic concrete frame building model, a value of 9 dB was found to be appropriate. $C_{1,2}$ is a frequency factor in Hz, controlling the width of the peak in the correction term at f_0. For the generic concrete frame building model, a value of 6 Hz was found to be appropriate.

The correction term C_2 accounting for damping loss at each storey is defined as:

$$C_2(f) = -\log\left(1 + \left(\frac{f}{C_{2,1}}\right) + \left(\frac{f}{C_{2,1}}\right)^4\right) \text{dB} \tag{4.8}$$

where $C_{2,1}$ is a frequency factor in Hz, acting like a cut-off frequency for a low-pass filter. For the generic concrete frame building model, a value of 50 Hz is suitable.

The mid-span floor vibration is then calculated using an additional correction term C_3, accounting for the modal effects of the floors and the vibration level at mid-span locations, defined as

$$C_3\left(f, f_{1,2,3}\right) = 3.3C_{3,1}\log\left(1 + e^{-\left(\frac{f-f_1}{C_{3,2}}\right)^2} + 0.5\,e^{-\left(\frac{f-f_2}{C_{3,2}}\right)^2} + 0.5\,e^{-\left(\frac{f-f_3}{C_{3,2}}\right)^2}\right) \text{dB} \tag{4.9}$$

The vibration level at mid-span locations is then

$$L_{v,mid}\left(f, n, f_{1,2,3}\right) = L_{v,col}(f, n) + C_3\left(f, n, f_{1,2,3}\right) \quad (4.10)$$

where $f_{1,2,3}$ are the first three modal frequencies of the floor (in Hz) that have a maximum at the centre of the floor span. These frequencies may be known from measurements or estimated from prediction models. Unpublished work by Lurcock has shown that for floors with symmetric boundary conditions and with an aspect ratio of *3:4*, $f_2 \sim 4f_1$ *and* $f_3 \sim 9f_1$.

$C_{3,1}$ is an amplitude factor representing the total amplification in decibels at f_1. A value of 8 dB may be appropriate in the absence of additional guidance from measurements or model data.

$C_{3,2}$ is a frequency factor in Hz, controlling the width of the peaks in the correction term at the modal frequencies. For the generic concrete frame building, a value of 7 Hz is suitable.

Petyt and Myrza[13] presented the results of computations of plates on point supports.

Auersch and Rücker, from BAM in Germany, developed a calculation method using transfer matrices, which for a one-dimensional system is a computationally efficient way of analysing the dynamic behaviour of a building. If excitation in the vertical direction is to be predicted, and the building response is considered in terms of vibration propagation up through walls or columns supporting floor plates, the system can be regarded as one-dimensional.

The transfer matrix method is a way of manipulating the system as a lumped-parameter two-port network such as the simple case illustrated in Figure 4.2. It involves a series of 2 × 2 square matrices in which the parameters associated with spring-like, mass-like and transmission line elements are entered.

The underlying equations are

$$\begin{aligned} P_s &= AP_r + BV_r \\ V_s &= CP_r + DV_r \end{aligned} \quad (4.11)$$

where P_s is the force and V_s is the velocity at the input to the system and P_r is the force and V_r is the velocity at the output from the system.

The matrix $\begin{bmatrix} A & B \\ C & D \end{bmatrix}$ is the transfer matrix. The input impedance is found be setting P_r to zero and it becomes B/D.

In the case of a series impedance, as in Figure 4.5, the transfer matrix is $\begin{bmatrix} 1 & Z \\ 0 & 1 \end{bmatrix}$

For a shunt impedance, as in Figure 4.6, the transfer matrix is $\begin{bmatrix} 1 & 0 \\ 1/Z & 1 \end{bmatrix}$, if the shunt network is connected to the receiving side of the series network, as in Figure 4.7, and the two corresponding matrices are multiplied.

The result is

$$\begin{bmatrix} 1 & 0 \\ 1/Z2 & 1 \end{bmatrix} \begin{bmatrix} 1 & Z1 \\ 0 & 1 \end{bmatrix} = \begin{bmatrix} 1 & Z1 \\ 1/Z2 & 1 + Z1/Z2 \end{bmatrix} \quad (4.12)$$

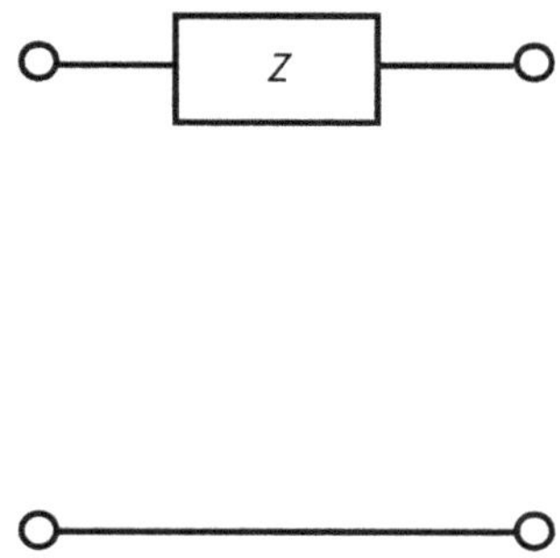

Figure 4.5 Series impedance.

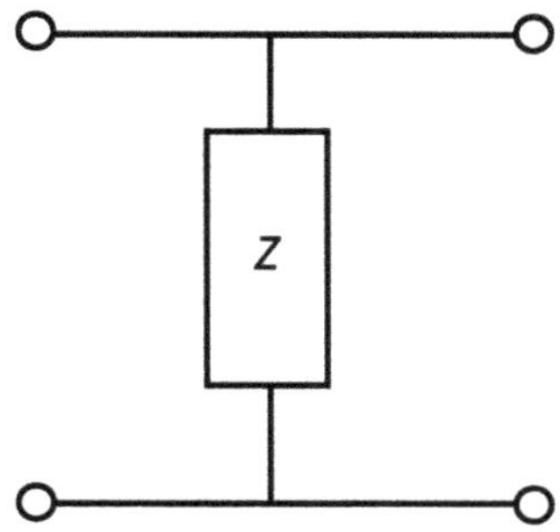

Figure 4.6 Shunt impedance.

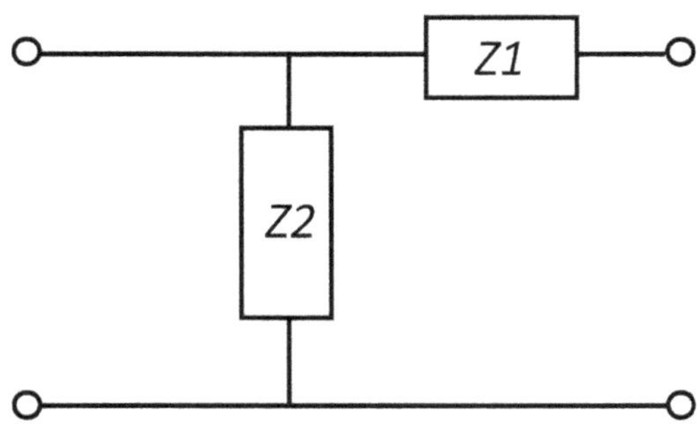

Figure 4.7 Shunt series network.

And the impedance at the source (left) side is $Z1/(1 + Z1/Z2) = 1/(1/Z1 + 1/Z2)$.

Any number of networks can be chained together in this way. If the impedance is a mass it is $i\omega m'$ and if a spring it is $k/i\omega$ where m' is the mass per unit area and k is the spring stiffness.

Where an impedance is an element, such as a column in a building, the transfer matrix is

$$\begin{bmatrix} \cos(kL) & Z_o\, i \sin(kL) \\ \dfrac{i}{Z_o}\sin(kL) & \cos(kL) \end{bmatrix} \tag{4.13}$$

where $k = \omega/c_o$ and $c_o = \sqrt{(E/\rho)}$ and L is the length of the column or wall and $Z_o = \rho_o c_o S_o$, with S being the cross-sectional area of the column.

The driving point impedance of an infinite plate is

$$Z_p = 8\sqrt{\frac{\rho_f E_f\left(1 + i\eta_f\right)h_f^4}{12\left(1 - \nu_f^2\right)}} \tag{4.14}$$

where ρ_f, h_f, ν_f, E_f and η_f are the material density, thickness, Poisson's ratio, Young's modulus and loss factor of the floor. Its transfer matrix is

$$\begin{bmatrix} 1 & Z_p \\ 0 & 1 \end{bmatrix} \tag{4.15}$$

Thus, if a tall building consisting of a stack of floor slabs and columns (or walls) is considered as a one-dimensional chain of impedances, these can be presented as transfer matrices and chained together. Auersch and Rücker[14] found that for a square foundation[15] on a homogeneous halfspace with Poisson's ratio = 0.33 the building should be treated as if supported on a base spring of stiffness $k' \approx 3.4\, G_s\sqrt{A_s}$ and damping $k'' \approx 1.6 A_s\sqrt{(G_s\rho_s)}$ where G_s is the soil shear modulus and ρ_s is the soil density.

The transfer matrix of the soil element is $\begin{bmatrix} 1 & 0 \\ 1/(k'/i\omega + c) & 1 \end{bmatrix}$

The coupling loss factor can then be determined first from the input impedance Z_b to the chain of transfer matrices, and then from the transmission coefficient

$$T_s = 1 - (Z_b/Z_s - 1)/(Z_b/Z_s + 1) \tag{4.16}$$

where, for incident p-waves $Z_s = \rho_s c_s S_s$ with ρ_s = soil density; c_s = p-wave velocity and S_s = foundation area.

Coupling loss factor is the ratio of the velocity of the building foundations to the velocity of the open ground surface, with no building present. It must be remembered that on an open ground surface there is velocity increase because the impedance of the air in the halfspace above the ground is close to zero, compared with that of the ground, so T_s becomes between 1.7 and 2, depending on the soil properties and the angle of incidence.

For an example building 15 m × 30 m, consisting of eight storeys of 250 mm thick concrete slabs supported on 600 mm diameter columns each

3.25 m high on a 3.6 m × 7.4 m grid, and supported on 12 m long piles, with external faces and internal partitions, a finite-difference simulation for excitation by an underground railway in a 5.6 m outside diameter tunnel 16 m below the ground surface in London Clay was carried out using the *FINDWAVE®* suite. The coupling loss factor was derived from the difference between vertical velocity in the slab below the columns and on the open ground surface. A comparison between the finite-difference result and the transfer matrix method is shown in Figure 4.8.

It is notable that the dip in the region of 40 Hz in both results is actually the natural frequency calculated from the mass of all the floor slabs above ground level and the spring constant of the columns supporting the first-floor slab. The loss factor at higher frequencies is controlled by the damping of the soil that was derived empirically by Auersch.[16]

In order to investigate the influence of building, foundation and soil parameters, the finite-difference model of the building, to which the results in Figure 4.8 apply and that is illustrated in Figure 4.9, was processed for a range of variations.

Figure 4.10 shows the effect of the soil assumptions on the coupling loss factor. The base case is a homogeneous halfspace of clay with shear modulus 0.735 GPa, density 1700 kg/m^3 and Poisson's ratio 0.4. This is compared with a 2 m deep layer of gravel above a 4 m layer of sand in place of the upper 6 m of clay.

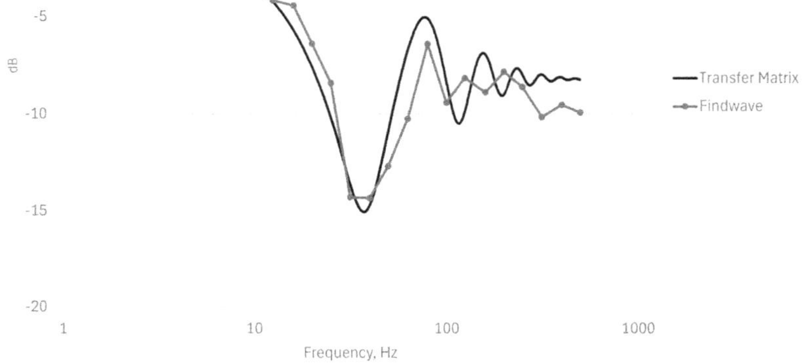

Figure 4.8 Comparison between finite-difference and transfer matrix calculations.

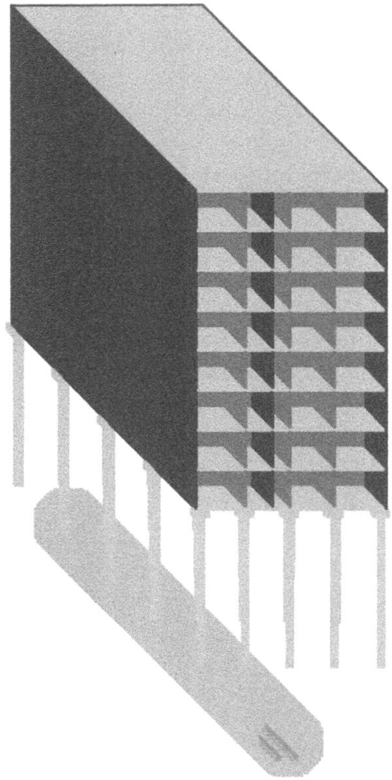

Figure 4.9 Finite-difference model of building above an underground railway (soil omitted for clarity).

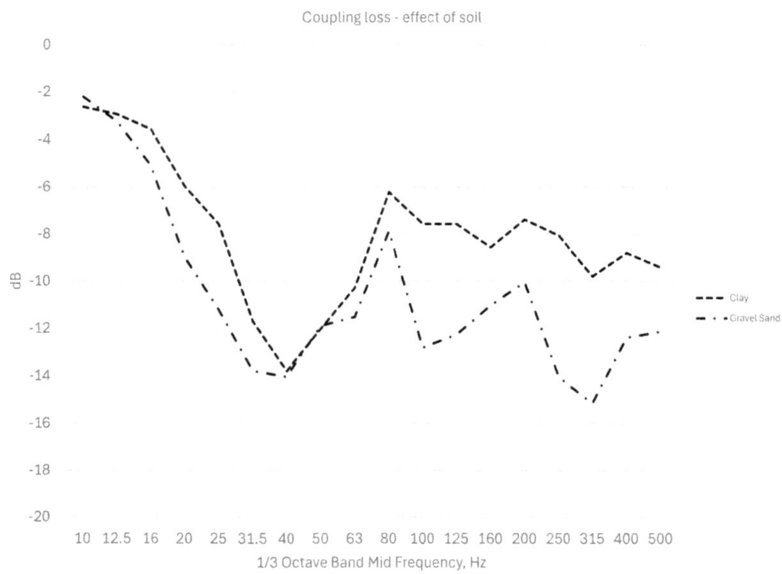

Figure 4.10 Effect of soil type.

NOTES

1. Degrande, G., Lombaert, G., Ntotsios, E., Thompson, D., Nélain, B., Bouvet, P., Grabau, S., Blaul, J., and Nuber, A. "State-of-the-art and concept of the vibration prediction tool." SILVARSTAR Deliverable D1.1, 28 May 2021.
2. Saurenman, H.J., Nelson, J.T., and Wilson, G.P., "Handbook of urban rail noise and vibration control." Prepared under contract to US DOT/Transportation Systems Center, Report UMTA-MA-06- 0099-82-2, February 1982.
3. Hanson, C.E., Towers, D.A., and Meister, L.D. "Transit noise and vibration impact assessment." FTA-VA-90-1003-06, U.S. Department of Transportation, May 2006.
4. Remington, P.J., Kurzweil, L.G., and Towers, D.A. "Low-frequency noise and vibration from trains." In *Transportation noise reference book*. P.M. Nelson (ed.), Butterworths, London, 1987.
5. Nelson, J.T., and Saurenman, H.J. "A prediction procedure for rail transportation groundborne noise and vibration." *Transportation Research Record*, 1143, 26–35, 1987.
6. Auersch, L. "Response to harmonic wave excitation of finite or infinite elastic plates on a homogeneous or layered half-space." *Computers and Geotechnics*, 51, 50–59, 2013.
7. Sanitate, G., and Talbot, J.P. "On the plate-like and layer-like response of slab foundations to ground-borne vibration." *Computers and Geotechnics*, 114, 103141, October 2019.
8. EN 15657: 2017, "Acoustic properties of building elements and of buildings - Laboratory measurement of structure-borne sound from building service equipment for all installation conditions."
9. Gibbs, B., and Villot, M. "Structure-borne sound in buildings: Advances in measurement and prediction methods." *Noise Control Engineering Journal*, 68(1), 1–20, 2020.
10. Moorhouse, A.T., and Gibbs, B.M. "Calculation of the mean and maximum mobility of concrete floors." *Applied Acoustics*, 45, 227–245, 1995.
11. Cremer, L., Heckl, M., and Ungar, E.E. *Structure-borne sound: Structural vibrations and sound radiation at audio frequencies*, 3rd ed., Springer, Berlin, 2005.
12. Lurcock, D., and Thompson, D. "A new empirical prediction approach for Groundborne vibration in buildings." *Proceedings of 24th International Congress on Sound and Vibration*, London, 2017.
13. Petyt, M., and Mirza, W.H. "Vibration of column-supported floor slabs." *Journal of Sound and Vibration*, 21, 355–364, 1972.
14. Auersch, L., and Rücker, W. "Praxisgerechtes Prognosemodell fur Erschutterungen – Einfache Rechenverfahren fur die Emission." Transmission und Immission, VDI-Berichte, 1754 Baudynamik, VDI-Verlag, Dusseldorf, 1–22, 2003.
15. Auersch, L. "Soil–structure interaction and damping by the soil – Effects of foundation groups, foundation flexibility, soil stiffness and layers." *Vibration* 8, 5, 2025. https://doi.org/10.3390/vibration8010005
16. Auersch, L. "Dynamic stiffness of foundations on inhomogeneous soils for a realistic prediction of the vertical building resonance." *Journal of Geotechnical & Geoenvironmental Engineering*, 134, 328–340, 2008.

Chapter 5

Wave types in structures

While groundborne noise is frequently generated underground and propagates through rock and/or soil by means of the variety of wave types discussed in Chapter 3, it is also frequently necessary to consider wave propagation in structures, either because the source is located in a structure or because the vibration is transmitted from the ground into a structure.

In order to comprehend the behaviour of structures in response to vibration excitation, it is important to understand some fundamental properties of finite structures.

BEAMS

Wave propagation in beams is important and interesting. Vibration propagates in beams in four ways. It can propagate longitudinally as a p-wave with a velocity $c_L = \sqrt{\frac{E}{\rho}}$ where E is Young's modulus and ρ is the mass density in kg per cubic metre. Thus, the phase speed is independent of frequency. It can propagate as a pure shear wave with velocity $c_S = \sqrt{\frac{G}{\rho}}$, where G is the shear modulus, again independent of frequency. It can propagate as a torsional wave with velocity $c_T = \sqrt{\frac{GK}{\rho I}}$ where K is a torsional constant and I is the polar moment of inertia of the section, once more independent of frequency. The torsional constant varies with shape and, for example is $\frac{1}{2}\pi r^4$ for a circular section; $2.25a^4$ for a square $2a \times 2a$. Tables of constants are given in reference books.[1]

The most important wave type is the bending wave which is dispersive, that is, frequency dependent, having a phase speed of $C_B \sqrt[4]{\omega^2 \frac{B}{\rho'}}$ where $B = Et^3/12$ is the bending stiffness for thickness t, and ρ' is the mass per unit length of the beam.

Where bending waves in beams become interesting is when they are supported by a resilient layer (such as a rail on resilient track support).

 DOI: 10.1201/9781003618775-5

The simple wave equation for bending waves, neglecting shear deformation and rotational bending effects (known as the Euler-Bernoulli theory) is

$EI\frac{\partial^4 u}{\partial x^4} + s'u + \rho'\frac{\partial^2 u}{\partial t^2} = 0$ where u is the vertical displacement and I is the second moment of area of the beam section and s' is the stiffness per unit length of the foundation. Noting that $\frac{\partial x}{\partial t} = c$ and $\frac{\partial}{\partial t} = i\omega$ this can be written as

$$c = \sqrt[4]{\frac{B}{\rho' - \frac{s'}{\omega^2}}}\sqrt{\omega} \tag{5.1}$$

The outstanding feature of this solution is that at the mass-spring natural frequency, when $\omega^2 = s/\rho'$, c becomes infinite. In physical terms, this means that a force applied to the beam causes the whole length of the beam to respond in phase. This frequency is known as the cut-on frequency, because above it the bending wave speed becomes finite.

In practice, the stiffness s' will include some damping, and become a complex number so that c remains finite, and complex, at the cut-on frequency. The effect of a complex phase speed is that the wave is attenuated with propagation along the beam.

The real (solid lines) and imaginary (dashed) values of c for a rail on a resilient support are given in Figure 5.1. The real part determines the velocity of propagation of a displacement along the beam. The imaginary part

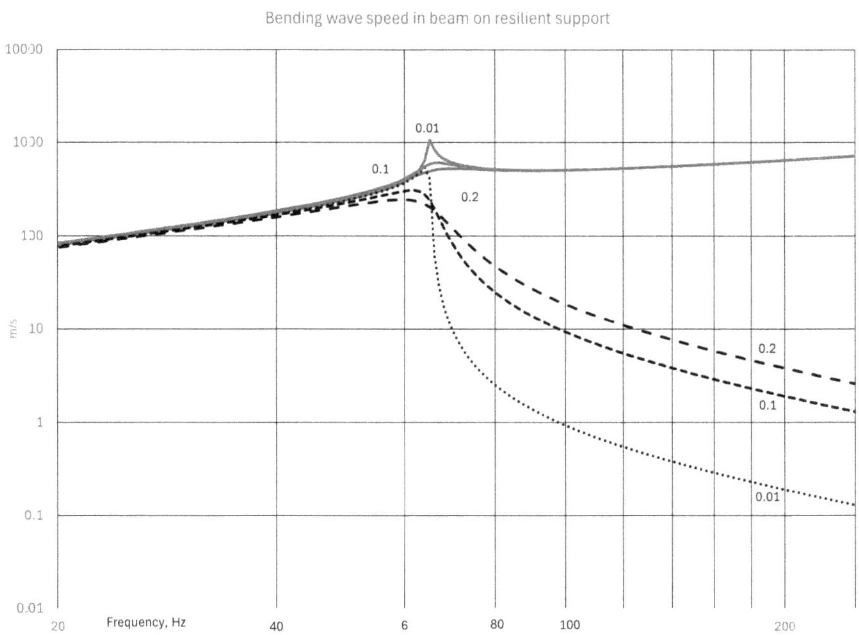

Figure 5.1 Real and imaginary parts of bending wave speed in a beam on a resilient support.

determines the decay rate. The wave propagates according to $e^{-i\omega x/c}$. The ratio $\frac{\omega}{c}$ is the wave number normally given the symbol $k = k' - ik''$, and the real part $k' = 2\pi/\lambda$ determines the wavelength λ and the imaginary part k'' determines the attenuation with distance given by $\Delta L = 20\ \log_{10}(\mathrm{e})k'' = 8.6859\,k''$ in dB per metre.

In the hypothetical case that there is no damping, then above the cut-on frequency k'' is zero, and there is no decay with distance. Below the cut-on frequency $k'' \approx -k'$ and the attenuation with distance is very large, such that the wave does not propagate. When a wave does not propagate for mathematical rather than damping reasons, it is said to be evanescent.

The bending wave speed is important in that for finite beams it determines the natural frequency of the beam. The most straightforward case is a beam with pinned-pinned support, and its natural frequencies are determined by the equation for c such that the lowest mode is at the frequency where the length between pinned supports is half a wavelength. This is the frequency of $\frac{\pi}{2l^2}\sqrt{\frac{B}{\rho'}}$ for a simply supported beam without a resilient support. For higher modes such as $n = 2, 3, 4$, in terms of angular frequency the equation is

$$\omega_n = \frac{(n\pi)^2}{l^2}\sqrt{\frac{B}{\rho'}} \tag{5.2}$$

where l is the distance between supports.

When a resilient support is included, that is, s has a finite value, the natural frequency is given by $\frac{1}{2\pi}\sqrt{\frac{\left(B\frac{\pi^4}{l^4} + s'\right)}{\rho'}}$. As before, when there is damping in the resilient support s' becomes complex. The inclusion of s' has the effect of increasing the natural frequency.

For beam supports other than pinned, the frequencies of the modes are higher. The most commonly encountered is clamped support at both ends, in which case the equation is

$$\omega_n = \frac{\pi^2(2n+1)^2}{4l^2}\sqrt{\frac{B}{\rho'}} \tag{5.3}$$

The same equation applies to free-free beams also, although such beams do not normally exist in the real world.

For a beam clamped at one end and free at the other, that is, a cantilever, the fundamental natural frequency is $\omega_1\left(\frac{3.5156}{l^2}\right)\sqrt{\frac{B}{\rho'}}$ and $\omega_2 = 6.268\,\omega_1$; $\omega_3 = 17.456\,\omega_1$; $\omega_4 = 34.375\,\omega_1$ and $\omega_5 = 56.79\,\omega_1$.

These equations relate to Euler-Bernoulli beams. When shear deformation and rotational effects are taken into account, it is necessary to turn to the wave equation solved by Timoshenko.[2] The effect is significant only when the bending wavelength is short compared to the dimensions of the beam. Allowing for the Timoshenko equation makes more than a 10% difference only if the wavelength λ is less than six times the thickness of the beam. In such cases the effect can be approximated by introducing a correction to the phase speed equation which becomes

$$C_B \sqrt[4]{\omega^2 \frac{B}{\rho'}} \left(1 - 3.6\left(\frac{h}{\lambda}\right)^2\right) \tag{5.4}$$

The driving point impedance of a thin beam is $0.5\rho' C_B(1 + i)$. In the case of a thick beam an approximation can be made by applying the Timoshenko correction of Equation (5.4).

PLATES

As with beams, the dynamic response of plates is dependent on the bending stiffness, mass and damping of the plate but also on the edge conditions. The edge of a plate may be free, simply supported (hinged) or clamped (encastré). Plates may have all four sides supported in the same way, of have a mixture of support types.

Plates occur in buildings as floors, and they may have any edge support conditions, and, in addition, may have intermediate supports in the form of columns.

The simplest case is that of hinged or simple support, because the natural frequencies of rectangular plates are capable of straightforward calculation. The natural frequencies are

$$f(m,n) = \frac{\pi}{2}\sqrt{\frac{B}{\rho h}}\left[\left(\frac{m}{a}\right)^2 + \left(\frac{n}{b}\right)^2\right] \tag{5.5}$$

where $B = \frac{Eh^3}{12(1-\nu^2)}$ in which E is Young's Modulus, ρ is the density; h is the plate thickness, ν is Poisson's Ratio; a and b are the length of the sides and m and n are integers.

While floor plates in structures may sometimes behave as if they are simply supported, in a reinforced concrete building the edges of concrete plates/slabs may be supported in a manner closer to the clamped condition than the simply supported condition. An exact solution for the natural frequencies of a rectangular clamped plate of dimensions $a \times b$ is not available, but an approximate formula for the natural frequencies for modes m, n is[3]

$$f = \frac{\lambda h \pi}{a^2}\sqrt{\left[\frac{E}{48\rho(1-\sigma^2)}\right]} \tag{5.6}$$

where $\lambda = \sqrt{G_x^4 + G_y^4 \frac{a^4}{b^4} + 2\frac{a^2}{b^2}\left(\sigma H_x H_y + (1-\sigma)J_x J_y\right)}$; h is the thickness and σ is Poisson's ratio.

The values of G and H are, for clamped edges (Table 5.1).
so since $H = J$ the result is not in this case dependent on Poisson's ratio.

For a plate with free edges (Table 5.2).

The natural frequencies for a plate clamped at all edges are very much higher than those for a plate simply supported at all edges, as can be seen from the table, for a concrete plate 200 mm thick, measuring 2.5 m by 4 m (Table 5.3).

An important conclusion is that a concrete slab with these properties and with free or clamped edges does not have a natural frequency within the frequency range normally significant for groundborne noise, but a simply supported plate does.

There is also a significant difference between the mode shapes of the simply supported and clamped cases, as illustrated in Figures 5.2–5.4. The mode shape function for the fundamental frequency for clamped edges is

$$\left(\cos\frac{2\pi x}{a} - 1\right)\left(\cos\frac{2\pi y}{b} - 1\right) \tag{5.7}$$

while that for simply supported edges is

$$\left(\sin\frac{\pi x}{a}\right)\left(\sin\frac{\pi y}{b}\right) \tag{5.8}$$

Table 5.1 Table of values for G, H and J in Equation (5.6) for plates with clamped edges

m or *n*	*G*	*H*	*J*
1	1.506	1.248	1.248
2 and above	$m + 0.5$	$(m + 0.5)^2\ [1-2/(m + 0.5)/\pi]$	$(m + 0.5)^2\ [1-2/(m + 0.5)/\pi]$

Table 5.2 Table of values for G, H and J in Equation (5.6) for plates with free edges

m or *n*	*G*	*H*	*J*
1	1.506	1.248	5.017
2 and above	$m + 0.5$	$(m + 0.5)^2[1 - 2/(m + 0.5)/\pi]$	$(m + 0.5)^2[1 - 6/(m + 0.5)/\pi]$

Table 5.3 Table of natural frequencies for free, clamped and simply supported edges

			n		
m	1	2	3	4	5
1	233	439	745	1156	1672
2	183	360	659	1070	1586
3	281	417	692	1090	1600
4	431	530	765	1137	1632
5	627	699	890	1225	1694
Free					
			n		
m	1	2	3	4	5
1	139	346	658	1075	1595
2	175	351	652	1063	1580
3	276	411	686	1085	1595
4	428	526	761	1133	1628
5	626	696	887	1221	1691
Clamped					
m	1	2	3	4	5
1	72	228	488	852	1319
2	133	289	549	912	1380
3	235	390	650	1014	1482
4	377	533	792	1156	1624
5	559	715	975	1339	1806
Simply supported					

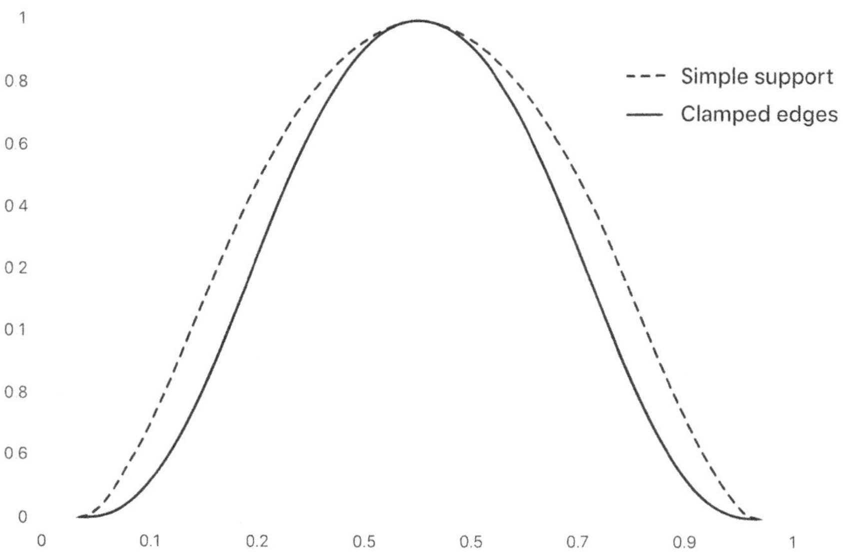

Figure 5.2 Mode shapes of simply supported and clamped plates, mode 1.

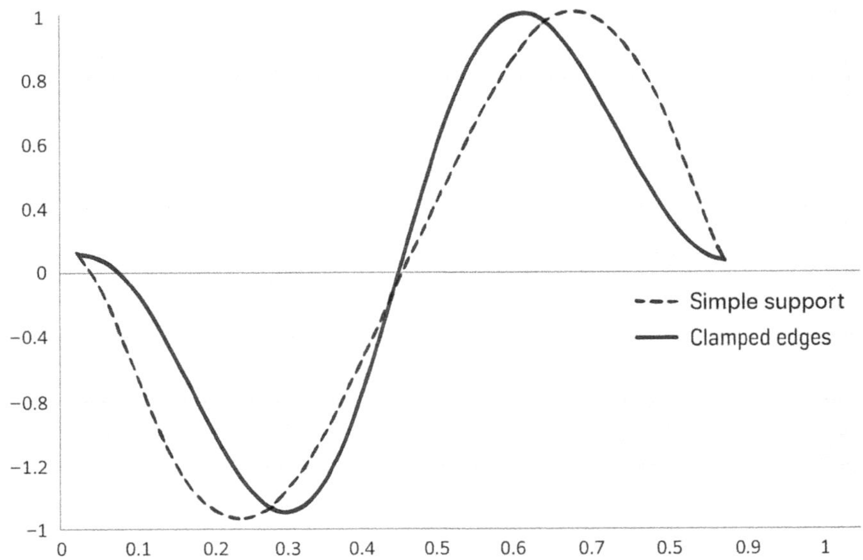

Figure 5.3 Mode shapes of simply supported and clamped plates, mode 2.

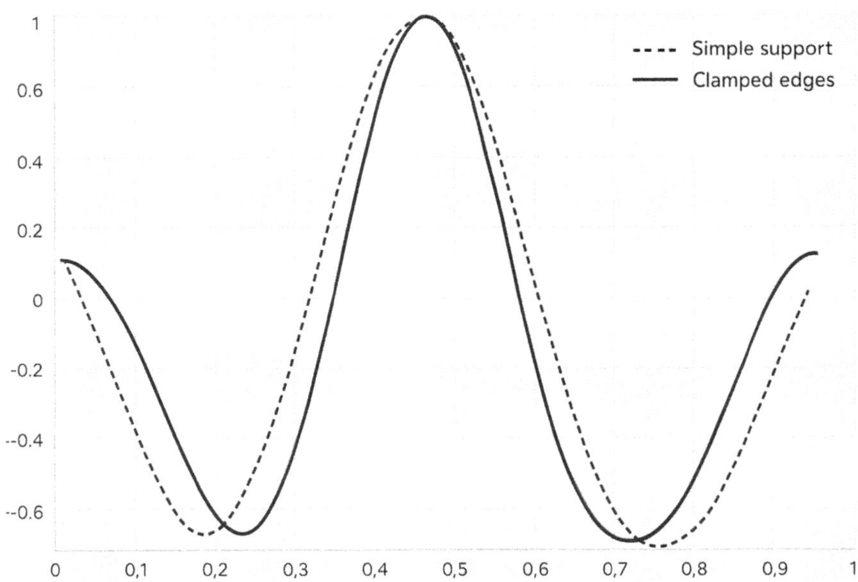

Figure 5.4 Mode shapes of simply supported and clamped plates, mode 3.

The significant difference relates to the edges, since when panels such as walls, floors and ceilings that are the boundaries of a room are vibrating, room modes have their antinodes in the corners.

While the mode shapes do not appear significantly different, it must be borne in mind that the frequencies at which they occur may be very different.

The driving-point impedance of an undamped simply supported rectangular plate of dimension x and y is[4]

$$Z_p = \frac{\rho}{4i\omega}\left(\sum_{n=1}^{\infty}\frac{sin^2\frac{n_1\pi x_0}{x}\,sin^2\frac{n_2\pi y_0}{y}}{\omega_n^2-\omega^2}\right)^{-1} \tag{5.9}$$

where ρ is the total mass of the plate and x_0 and y_0 are the distances of the driving point form the plate origin. Each eigenfrequency $\omega_n = 2\pi f(n_1, n_2)$ is obtained from Equation (5.5).

The undamped case is unrealistic as it yields zero impedance for each eigenfrequency. Damping may be included by making E complex in Equation (5.3), and the effect is illustrated in Figure 5.5.

As discussed further in Chapter 11, when a plate such as a floor slab has a resiliently supported layer attached to it, the impedance of the resilient layer affects the dynamic behaviour of the supported layer. At the mass-spring

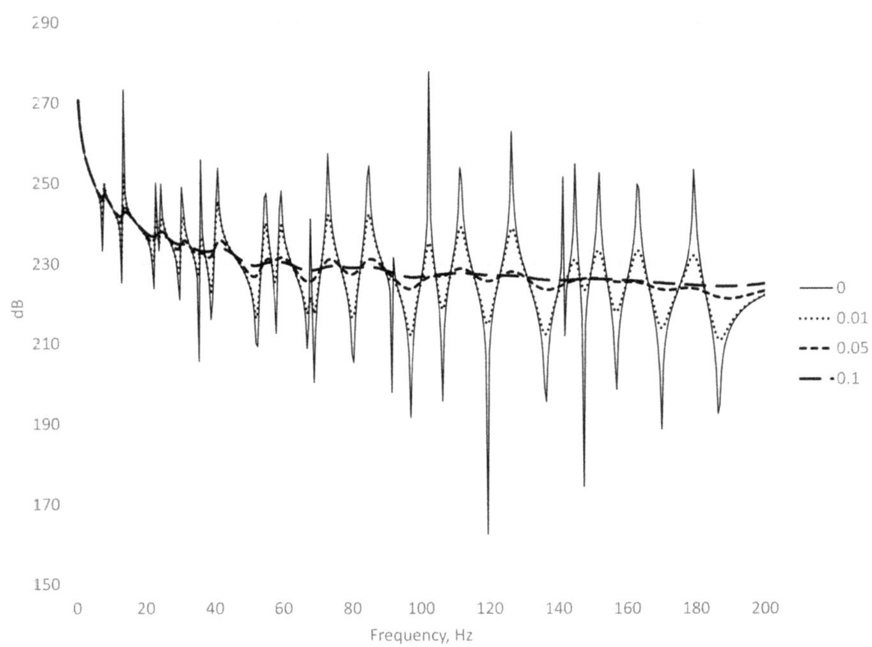

Figure 5.5 The driving point impedance of a simply supported plate excited near a corner, with loss factors from 0 to 0.1.

natural frequency of the added system, the same effect occurs with a resiliently supported beam in that the bending wave speed tends to infinity, which means that all of the supported panel moves vertically in-phase, and the lowest plate mode in the supported panel is increased, The equation for the panel modes of the resiliently supported layer becomes

$$\omega_{(m,n)} = \sqrt{\frac{B}{\rho''}\left[\left(\frac{n\pi}{l_x}\right)^2 + \left(\frac{m\pi}{l_y}\right)^2\right]^2 + \frac{s''}{\rho''}} \tag{5.10}$$

where $B = \frac{Et^3}{12(1-\sigma^2)}$; E is Young's modulus; t is the panel thickness and σ is Poisson's ratio, ρ'' is the mass per unit area of the panel and s'' is the stiffness per unit area of the resiliently supporting layer; l_x and l_y are the panel dimensions and n and m are any integers. The stiffness of the supporting layer will be a complex number because of its inherent damping, and indeed damping will control the amplitude of peaks in the spectrum of panel modes. The stiffness of the supporting layer will have damping properties, in which case s'' is a complex number.

WAVE TRANSMISSION THROUGH STRUCTURES

When waves in beams and plates travel through structures, they will meet discontinuities such as changes of section and junctions. Reflection occurs at each discontinuity, and it is possible to calculate the transmissions and reflection coefficients based on the properties of the discontinuity[4] for both longitudinal and bending waves. For longitudinal waves in a beam the reflection coefficient is $r = (Z_1 - Z_2)/(Z_1 + Z_2)$ where $Z_1 = \rho_1 c_{L1} S_1$ and $Z_2 = \rho_2 c_{L2} S_2$, ρ being the density, c_L being the longitudinal wave speed and S the cross-sectional area in the structure either side of the discontinuity. In many cases $\rho_1 = \rho_2$ and $c_{L1} = c_{L2}$, so that the reflection coefficient is merely a function of the cross-sectional areas S_1 and S_2. For bending waves the situation is much more complex, with a reflected wave r and an evanescent reflected wave r_i. These are a function of the ratio of bending wavelengths either side of the discontinuity, Ψ, so that $r = \frac{2\psi(1-\kappa^2)-i\kappa(1-\psi)^2}{\kappa(1+\psi)^2 + 2\psi(1+\kappa^2)}$ and $r_i = \frac{\kappa(1-\psi^2)-i\kappa(1-\psi)^2}{\kappa(1+\psi)^2 + 2\psi(1+\kappa^2)}$.

While for longitudinal waves the transmission coefficient is simply 1+r, for bending waves there are two coefficients $t = \frac{2(1+\kappa)(1+\psi)}{\kappa(1+\psi)^2 + 2\psi(1+\kappa^2)}$ and $t_i = \frac{2(1-\psi)-i2\kappa(1-\psi)}{\kappa(1+\psi)^2 + 2\psi(1+\kappa^2)}$.

Reflection and transmission also occur at corners and at junctions between beams and plates. and the equations are set out in reference 4.

It has to be borne in mind that there are usually several discontinuities along the transmission path of a longitudinal or bending wave, and a reflected wave from one discontinuity will travel back to a previous discontinuity. In such cases where there is a chain of impedance mismatches, it is possible to invoke the transfer matrix method to calculate the input impedance to a chain of impedance mismatches, as explained in Chapter 4. However, in a three-dimensional array of discontinuities, such as a reinforced concrete building with columns and floor plates, it is more appropriate to use numerical methods for simulating the dynamic response of the system, as described in Chapter 8.

DAMPING

The concept of damping is referred to in several chapters of this book. While it is an important topic, it can also be a confusing one, as damping is quantified in more than one way, using more than one descriptor. It is, therefore, worthwhile bringing together the different damping descriptors and their relationships with one another.

Damping occurs for a number of reasons. Perhaps the most commonly discussed form is viscous damping that arises when fluid in a medium is set into motion and energy is removed from the motion as a result of viscosity in the medium. Viscous damping is frequency dependent.

Damping also occurs due to hysteresis, which is a lag between stress and strain in an oscillating system. Hysteretic damping is independent of frequency.

An allied effect is due to soil plasticity. This is referred to as Masing behaviour in Chapter 6, through which changes in stiffness and hysteretic loading-unloading introduce damping. Although numerical modelling of plasticity effects is possible,[5] significant effects due to plasticity are normally found only in cases of much greater strain than is the case for groundborne noise.

Friction at the interface between moving surfaces, particles or objects causes damping, and this is also known as Coulomb damping. In this case the damping is not proportional to the magnitude of velocity, as is the case in viscous damping.

In the consideration of groundborne noise, the most directly relevant descriptor is η, which emerges directly from the consideration of the elasticity of a medium such as soil in terms of a complex modulus, so that η is then the ratio of the imaginary part of the complex modulus to the real part. A complex modulus can also be written in terms of modulus and argument, where the modulus is expressed as the square root of the sum of the squares of the real and imaginary part, and the argument is the angle whose tangent is the imaginary part divided by the real part. It is common practice to assign the symbol delta to this angle, and then η is $\tan \delta$. In some publications "tan delta" is used in place of η.

Damping is also key in situations where the topic of interest is not energy dissipation in the transmission of wave motion through a medium such as soil, but in the height of the peak in a resonant system. While the simple case is a mass-on-a-spring, a related case is a resonant structure, such as a floor panel in a building. Damping will determine the amplitude of the peak in the resonant response. Three further parameters are used to describe damping in the case of a mass-spring system. The first is the logarithmic decrement, confusingly also usually given the symbol δ, which is the natural logarithm of the ratio of the amplitudes of successive peaks in the decay curve that results from an impulse excitation, or the instantaneous removal of strain as in a step function, divided by π. The exact equation is $\delta = 2\pi\zeta/\sqrt{(1-\zeta^2)}$, which for small values of zeta is approximately $2\pi\zeta$ so η is $\ln(x(t_0)/x(t_N))/N/\pi$. Another parameter is the ratio of the damping rate to the level of critical damping. Critical damping occurs when there is just enough damping for the system to return to its equilibrium state with no overshoot. It is typically given the symbol C_c and C/C_c is given the symbol ζ, which the equal to $\eta/2$. Yet another parameter is amplification at resonance, usually given the symbol Q and named Quality Factor. Q is $1/\eta$ or $1/2\zeta$. Lastly, damping can be expressed in terms of the 3 dB down point, being the frequency at which the power in the signal is half that at its resonant frequency. Q is $f_0/\Delta f$ where f_0 is the resonant frequency and Δf is the bandwidth between the two 3 dB down points.

In summary, all these measures of damping can be stated as

$$\eta = 2\zeta = 1/Q = 2C/C_c = \delta/\pi$$

With regard to damping in a propagation medium where the mechanism is hysteresis, damping can be expressed in terms of the measured plot of stress versus strain, or force versus displacement, known as the hysteresis loop. The area within the hysteresis loop is divided by the maximum strain energy. The area of the hysteresis loop is denoted ΔW, and half the product of the maximum strain multiplied by the maximum stress is denoted as W. The damping D is then $1/4\pi\Delta W/W$.

SIMULATION OF DAMPING IN NUMERICAL MODELS

Rayleigh damping

Numerical models, particularly those using Finite Element Analysis (FEA), often implement a damping model known as Rayleigh damping. It is an approximation to viscous damping that allows modelling of the energy dissipation in the material due to internal friction, assuming it is proportional to strain or rate of deformation.

The damping force term is assumed to be proportional to the deformation rate as in the dynamic equation for a one degree of freedom system with mass, damping coefficient and spring stiffness.

The Rayleigh model approximates the damping coefficient as a linear combination of mass and stiffness:

$$c = \alpha k + \beta m$$

where α is stiffness-proportional damping coefficient and β is mass-proportional damping. The equation of motion then becomes:

$$\ddot{x} + 2\zeta\omega_n\dot{x} + \omega_n^2 x = \frac{F(t)}{m} \tag{5.11}$$

where $\omega_n^2 = \frac{k}{m}$; $\zeta = \frac{C}{C_c}$.

The Rayleigh damping coefficients are related to the damping ratio:

$$\zeta = \frac{1}{2m\omega_n}(\alpha k + \beta m) \tag{5.12}$$

The Rayleigh model can reproduce three cases. The first is obtained by setting $\alpha = 0$ so that $\zeta = \beta/2\omega_n$ and damping is inversely proportional to natural frequency. The second is obtained by setting $\beta = 0$ so that damping is proportional to the stiffness and $\zeta = 1/2\ \alpha\omega_n$. In this case damping is directly proportional to natural frequency.

Usually in FEA models damping is represented with finite values for α and β so that $\zeta = 1/2(\alpha\omega_n + \beta/\omega_n)$ and the frequency-dependence of ζ is such that it increases at frequencies below ω_1 and above ω_2 and is at its minimum between ω_1 and ω_2.

Damping of this kind rarely occurs in the real world, and its usefulness in numerical modelling is primarily to enable the desired value to be achieved in the range between ω_1 and ω_2, which are selected to cover the principal frequency responses in the model, at the expense of inaccurate damping at higher and lower frequencies.

NOTES

1. Young, W.C., *Roark's formulas for stress & strain*, McGraw Hill Book Company, New York, 1989.
2. Timoshenko, S.P. "On the correction for shear of the differential equation for transverse vibrations of prismatic bars." *The London, Edinburgh, and Dublin Philosophical Magazine and Journal of Science, Series* 6, 41, 744–746, 1921.

3. Warburton, G.B. "The vibration of rectangular plates." *Proceedings of the Institution of Mechanical Engineers, Series A*, 168(12), 371–384, 1954.
4. Cremer, L., Heckl, M., and Petersson, B.A.T., *Structure-borne sound*, 3rd ed., Springer, Berlin, 295, 2005.
5. Whyte, S.A., Burd, H.J., Martin, C.M., and Rattley, M.J. "Formulation and implementation of a practical multi- surface soil plasticity model." *Computers and Geotechnics*, 117, 103092, 2020.

Chapter 6

Soil dynamic properties

Groundborne noise (which is vibration at frequencies within the audible range) propagates as wave motion through shearing of the soil and through compression of the soil and/or water in the soil. While these are the only types of wave motion in an infinitely large homogeneous medium, when discontinuities appear, such as interfaces between two kinds of soil, or between soil and air (at the ground surface), the shear and compression waves combine in a way which gives rise to other types of wave motion. Ground surface waves are known as Rayleigh waves, or Love waves when motion is in the plane of the ground surface and perpendicular to the direction of propagation. Waves at interfaces are known as Stoneley waves, and waves in layers are known as Lamb waves.

While density does not vary significantly (except as a result of water saturation), shear modulus is strain dependent and stress dependent (shear modulus varies with depth below ground because of the consequent variation of the load of the soil above, known as confining stress). Poisson's ratio is dependent on water saturation, and to some extent on strain. Strain is the displacement of a sample as a result of stress, which may be force or pressure. Shear modulus can be frequency dependent, and loss factor usually has some dependence on frequency.

Wave amplitudes decay with distance due to geometric spreading, that is, the waves expand as they travel away, and, therefore, the energy they transport is diluted over a larger area, resulting in lower amplitudes at any particular point. Wave amplitudes also decay due to conversion of vibration energy into heat, a process known as damping. Waves are reflected at interfaces between different media, and the completeness of the reflection depends on the contrast between the modulus and the density of each of the media. The amount by which transmitted wave amplitudes are reduced by reflections at layers depends also on the angle of incidence of the wave, and the effect of other layers nearby. At reflection, conversion between wave types occurs, and, in particular, a compression wave may be reflected as a compression wave and a shear wave. Since shear waves are more rapidly attenuated due to damping, reflections cause increased attenuation.

DOI: 10.1201/9781003618775-6

Damping is measured and expressed in many ways, all inter-related. In the propagation of vibration through soil, it can be expressed in terms of the excess loss, over and above that due to geometric effects, in decibels per second (converted into decibels per metre, with knowledge of the wave speed of the soil for the wave type concerned). This, in turn, is dependent on a quantity known as loss factor, which fundamentally is a measure of the extent to which force keeps pace with displacement (i.e., stress is in phase with strain), when a sample is cyclically compressed and decompressed. In the case of a loss-free sample, the force induced by compressing it is simply dependent on the amount by which it is compressed (or decompressed). In a sample with a finite loss factor, the stress will be partly dependent on the strain, and partly on the velocity. Since velocity is zero at maximum and minimum strain, it follows that stress will be maximum at some point other than maximum or minimum strain, and the extent to which strain lags or leads stress gives loss factor, η, measured as the tangent of the phase angle between the two. Many other measures of damping exist, including damping ratio, which is half loss factor, and Q which is the inverse of loss factor.

Soils are non-linear at large strains. As Figure 6.5 shows, a positive force which compresses a sample is associated with the same strain as a negative force during decompression. This effect, known as Masing behaviour, causes significant loss factors (e.g., 0.3) in soils at large strains. The non-linearity is caused by the dependence of shear modulus, G, on strain. The value of G for infinitesimally small strain is known as G_{max}.

In order to be able to interpret much of the literature in the context of groundborne noise, it is helpful to consider the likely range of strains that occur. A common source of vibration and groundborne noise is the operation of railways. Typically, rail vibration velocity is about 140 dB re 1 nanometre/second (nm/s). Invert vibration will tend to be about 20 dB below this, and, assuming a pessimistic case that the compressional wave speed is 600 m/s (e.g., gravel), the strain, given by v/c where v is the particle velocity, is 0.00017%. This is the strain due to vibration. There will be locally larger strain, due to the dead load of the tunnel, that could be of the order of 0.1%,[1] but which rapidly decays with distance, particularly to the side of the tunnel.

However, only in earthquakes, the near field of pile driving, blasting, large out-of balance machinery or vibration of a large structure such as an offshore oil rig due to wave and wind action, (and in some geotechnical testing), does large strain vibration occur. In the propagation of groundborne noise from underground railways, the strains are infinitesimally small, and the damping effect due to Masing behaviour is very small (e.g., 0.01).

Other contributors to loss factor are viscosity, if fluid movement is involved, and friction (particularly in the case of sands).

In some soils, loss factor is not strongly dependent on frequency; but this is not always the case. However, if loss factor is converted into decay in terms of dB per metre, the result, for a constant loss factor is proportional to frequency. This can give very high losses in terms of dB per metre at "rumble" frequencies such as 100 Hz, and it is, therefore, of great importance that the loss factor is not over-estimated. If, for example, a loss factor of 0.3 were true for small strains the loss in dB per metre at 100 Hz can be of the order of 1 dB per metre. The relationship is

$$\Delta L = \frac{4.34\omega\eta x}{c} \tag{6.1}$$

where $\omega = 2\pi f$ and f is frequency, and c is the wave speed concerned. This highlights the importance of correct assumptions about wave speed (in turn dependent on modulus and density), as well as about loss factor.

Reflections between layers, as mentioned above, are dependent on the mismatch between soil dynamic properties at the interface between the layers. The important parameter is the product of the wave speed and the density and is called impedance. Since wave speed is the square root of modulus over density, impedance is the square root of modulus times density.

Damping can be conveniently expressed by using complex algebra, that is, terms which have two parts known as real and imaginary. When considering acoustical phenomena in terms of individual frequencies, that is, single sine waves, the waves have two properties, amplitude and phase (indicating the time at which the wave reaches its crest). These properties can be expressed as a magnitude (the height of the crest) and an argument (the phase angle obtained by expressing the position of the crest along a scale in which one complete cycle is 180° or 2π radians). The same information expressed in complex form has a first term (called "real") which is the magnitude multiplied by the cosine of the argument and a second term (called "imaginary") which is the magnitude multiplied by the sine of the argument. Effectively this breaks the wave down into a part which has its maximum at time zero and at multiples of $1/f$ where f is the frequency, and another wave of the same frequency but different amplitude that is 90° out of phase. To signify the 90° out-of-phase, the imaginary part carries the symbol i, which means that when the imaginary part is squared in calculations, the result is negative. The great advantage of complex notation in acoustics is that the ratio of the imaginary part to the real part is the loss factor, and single frequency waves can be expressed as exponents using Euler's identity $e^{ix} \equiv \cos(x) + i\sin(x)$. Not only is algebraic manipulation much simpler, but if x itself is complex, that is, $x = \alpha + i\beta$ then e^{ix} becomes $e^{i\alpha}$ multipled by $e^{-\beta}$. The first of these terms is the same sine and cosine function, but the second decreases in amplitude with increasing x. In the time domain, $x = \omega t$ where t is time, so that the effect of the $e^{-\omega t}$ is exponential decay of an oscillation with time, that is, damping.

METHODS FOR ESTIMATION OF DYNAMIC PROPERTIES OF THE GROUND

Stratification and classification of the ground

In estimating the dynamic properties of the ground it is first of all necessary to identify the stratification and classification of the ground. For this purpose it is essential to obtain basic geotechnical and geological site investigation data (boring logs, soil types, stratification and water table). Existing investigations for geotechnical design often provide useful index parameters. Supplementary investigations may be necessary for vibration/noise-specific parameters.

Soils versus rock

Soils and rocks differ distinctly in dynamic behaviour; and, therefore, different methods apply. Transitional cases (e.g., residual soils) exist with ambiguous classification.

Numerous empirical/semi-empirical correlations exist to estimate dynamic moduli and wave speeds from geotechnical index parameters including porosity, plasticity index, undrained shear strength, Standard Penetration Testing (SPT) and Cone Penetration Testing (CPT) values. For cohesionless soils shear modulus relates to void ratio and effective stress. For clays shear modulus can be estimated from undrained shear strength and plasticity index. Rock mass quality Qc correlates with p-wave velocity. Dilatational wave speed may be estimated from shear wave speed and Poisson's ratio, with adjustments for saturation.

w?>CPT and SPT results can be used to estimate dynamic parameters via empirical correlations. CPT cone resistance and corrected tip resistance are linked to shear modulus. SPT blow counts are related to shear wave velocity and modulus. These methods provide rough estimates suitable for early screening.

More reliable data can be obtained by direct seismic methods measuring wave propagation and attenuation, including surface wave methods (non-intrusive, average properties over larger volumes, depth range ~30–50 m), seismic CPT (S-CPT) (down-hole seismic measurements combined with CPT), down-hole and up-hole seismic logging, cross-hole seismic testing (high accuracy, requires multiple boreholes) and other methods including (refraction, multi-channel reflection, resistivity for stratigraphy). Consideration must be given to stress dependency of parameters.

Surface wave measurements

Surface waves (Rayleigh waves) are dispersive when the soil properties vary with depth; phase velocity varies with frequency in accordance with the vertical shear wave velocity profile. Measurement involves generating surface waves, recording propagation, processing dispersion curves and inversion to derive Vs-depth profiles. This method requires adequate seismic sources, sensor arrays, and advanced processing including Multichannel Analysis of Surface Waves (MASW), and Spectral Analysis of Surface Waves (SASW) and is well suited for broad site characterisation.

Down-hole and up-hole measurements (including S-CPT)

This involves single borehole measurements where source and receivers are located at different depths. Seismic Cone Penetration Testing (S-CPT) combines seismic sensors with CPT probes for efficient measurement in soft to medium soils. Up-hole is the reverse configuration. The method is useful for obtaining detailed profiles of shear wave velocity V_s and p-wave velocity V_p but has limited penetration (30–60 m).

Cross-hole measurements

This method uses multiple parallel boreholes with sources and receivers at matching depths. It provides accurate measurement of wave speeds and attenuation. It requires borehole drilling and alignment and is the most accurate, but more costly and intrusive method.

Other measurements

Refraction seismic surveys provide p-wave velocities and stratigraphy. Multi-channel shear wave reflection surveys are used for detailed shallow imaging. Geo-electric methods (e.g., resistivity tomography) give indirect stratigraphic information. Combined geophysical and geotechnical approaches are, however, recommended.

Other in-situ methods

Dynamic plate load tests, vane tests and others can provide high-strain properties. Stress dependency must be accounted for. These methods have limited applicability for low-strain vibration issues.

Laboratory measurement of dynamic ground parameters

Laboratory tests on samples allow controlled measurement of dynamic moduli and damping over strain ranges, although there can be issues with sample disturbance and representativeness of samples. The methods include piezoelectric bender elements (V_s measurement under low strain), resonant column tests (that measure moduli and damping including non-linear behaviour) and cyclic triaxial, direct simple shear (DSS), and torsional shear tests (that cover higher strain range). Laboratory testing provides additional insight into non-linear soil behaviour and loss factors but is less representative of in-situ conditions.

THEORY

The topics that have been researched most extensively, in the contexts of the acoustics of the ocean floor (for defence purposes), earthquake and seismic research, construction vibration and blasting, are wave speeds, shear modulus and damping. Literature on the propagation of vibration from railways tends to give data only on overall distance attenuation, which is of little use if the mechanisms that contribute to distance attenuation are not disaggregated. For present purposes, data that do not relate damping to frequency, and do not relate moduli and wave speeds to variables such as depth below ground, are of limited value. Extensive data exist for the ocean floor and are of direct relevance, except that the soil types tend to be silts, clays, sands and rock, with little data on gravel. The dependence of shear moduli on depth below the ocean floor cannot be directly converted to depth below the ground surface, or to dry materials.

WAVE SPEEDS

Lew and Campbell[2] give a relationship between shear wave velocity and depth of overburden. They found in a previous study that the relationship of shear-wave velocity with depth could be adequately represented by the expression $c_s = K(d + e)^n$ where c_s is shear wave velocity; d is depth; K, q, e and n are constants dependent upon the geotechnical classification. Term e accounts for non-zero values of c_s at the ground surface. They give a table of near-surface velocities derived from statistical analyses of c_s and c_p values of the surface deposits and a few near-surface deposits. These are given in Table 6.1.

Table 6.1 Lew and Campbell's values for near-surface velocities

Soil		*Shear wave velocity, c_s, m/s mean and standard deviation*	*Compressional wave velocity, c_p, m/s mean and standard deviation*
Soft natural soil <10% gravel and cobbles –very recent floodplain, lake, swamp, estuarine, and delta deposits and hydraulic fill soils; may contain organics, less than 1600 kg/m^3 dry density	Mean	161	356
Saturated Holocene age soils, between 1400 and 1750 kg/m^3	Standard deviation	18	98
Very low-density soils; primarily fine-grained, dry density less than 1400 kg/m^3			
Soft to moderately soft clays and clayey silts; generally dark grey to black, dry density less than 1600 kg/m^3 at the surface and less than 1300 kg/m^3 below 3 m.			

(Continued)

Table 6.1 (Continued)

Soil		*Shear wave velocity,* c_s*, m/s mean and standard deviation*	*Compressional wave velocity,* c_p*, m/s mean and standard deviation*
Soft clay (for definition of soft see "soft natural soil") more than 3 m depth	Mean	94	332
	Standard deviation	27	—
Soft clay (for definition of soft see "soft natural soil") 3 m to 30 m depth	Mean	192	—
	Standard deviation	21	—
Intermediate natural soil < 10% gravel and cobbles, unsaturated Holocene age soils of moderate density; may contain moderate amounts of gravel at depths below 10 to 15 m, dry density 1400 to 1750 kg/m^3	Mean	214	432
Saturated, uncemented Pleistocene and Eocene age soils; may contain moderate amounts of gravel at depths below 3 to 6 m, dry density 1500 to 1840 kg/m^3 at 3 to 5 m depth	Standard deviation	40	111
Low density Pleistocene and Eocene age soils at the surface, dry density 1400 to 1750 kg/m^3			
Firm natural soil < 10% gravel and cobbles, high density Holocene age soils, dry density greater than 1500 kg/m^3	Mean	266	510
Unsaturated, uncemented Pleistocene and Eocene age soils of moderate density; may contain moderate amounts of gravel at depths below 3 to 6 m, dry density 1600 to 1840 kg/m^3	Standard deviation	46	122
High density Pleistocene and Eocene age soils at the surface, dry density greater than 1875 kg/m^3			

(*Continued*)

Table 6.1 (Continued)

Soil		*Shear wave velocity, c_s, m/s mean and standard deviation*	*Compressional wave velocity, c_p, m/s mean and standard deviation*
Non-engineered fill	Mean	158	404
	Standard deviation	17	74
Engineered fill	Mean	264	488
	Standard deviation	—	—
10–50% gravel, depth = 0 m	Mean	317	686
	Standard deviation	—	—
10–50% gravel, depth = 1.5–18 m	Mean	398	748
	Standard deviation	57	71
10–50% gravel with cobbles, 50% gravel, depth = 1.5–18 m	Mean	487	989
	Standard deviation	125	222
Saturated soil	Mean		1605
	Standard deviation		133

Table 6.2 Coefficients for shear wave velocity estimation

Soil classification	*K*	*n*	*e*	*Standard deviation in ln C_s expressed as multiplier of C_s*
Soft natural soils	105.88	0.385	1.624	1.211
Intermediate and saturated firm natural soils	128.7	0.402	1.597	1.160
Firm soils	222.2	0.280	0.165	1.156

Lew and Campbell's values for K, n and e for shear waves, using the soil classifications given for this reference in Table 6.1, are given in Table 6.2.

The corresponding values for compression waves are given in Table 6.3.

A further set of coefficients is given to combine shear wave velocities and compression wave velocities through a multiplier m to be applied to the shear wave velocity, to arrive at compression wave velocity, as shown in Table 6.4.

Figures 6.1 and 6.2 use the coefficients from Table 6.4. for compressional waves.

Hamilton[3] quotes a regression equation by Ohta and Goto for shear wave velocity versus depth in water-saturated silt-clays and turbidites on the ocean

Table 6.3 Coefficients for compression wave velocity wave estimation

Soil classification	*K*	*n*	e	*Standard deviation in C_s expressed as a multiplier of C_s*
Soft natural soils	Not	sufficient	data	
Intermediate and saturated firm natural soils	329	0.305	0.427	1.195
Firm soils	221	0.288	0.287	1.212

Table 6.4 Coefficients and multipliers for shear and compression wave estimation

Soil classification	*K*	*n*	e	*Standard deviation in ln C_s expressed as a multiplier of C_s*
Soft natural soils	216	0.375	1.448	1.222
Intermediate and saturated firm natural soils	277	0.370	1.186	1.17
Firm soils	421	0.284	0.223	1.18

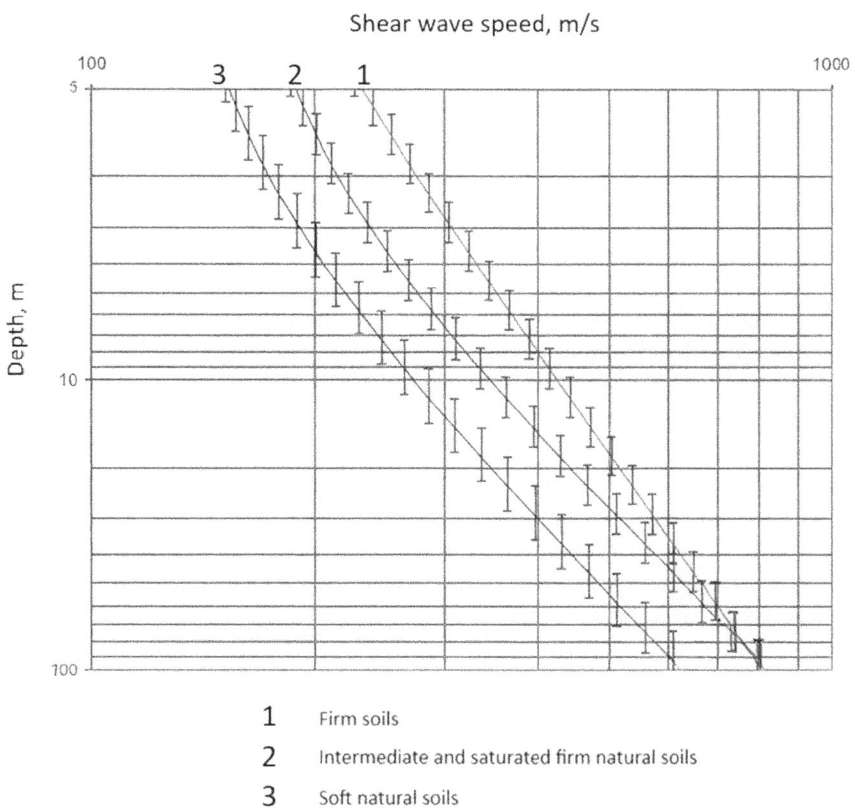

Figure 6.1 Shear wave speed of soils as a function of depth.

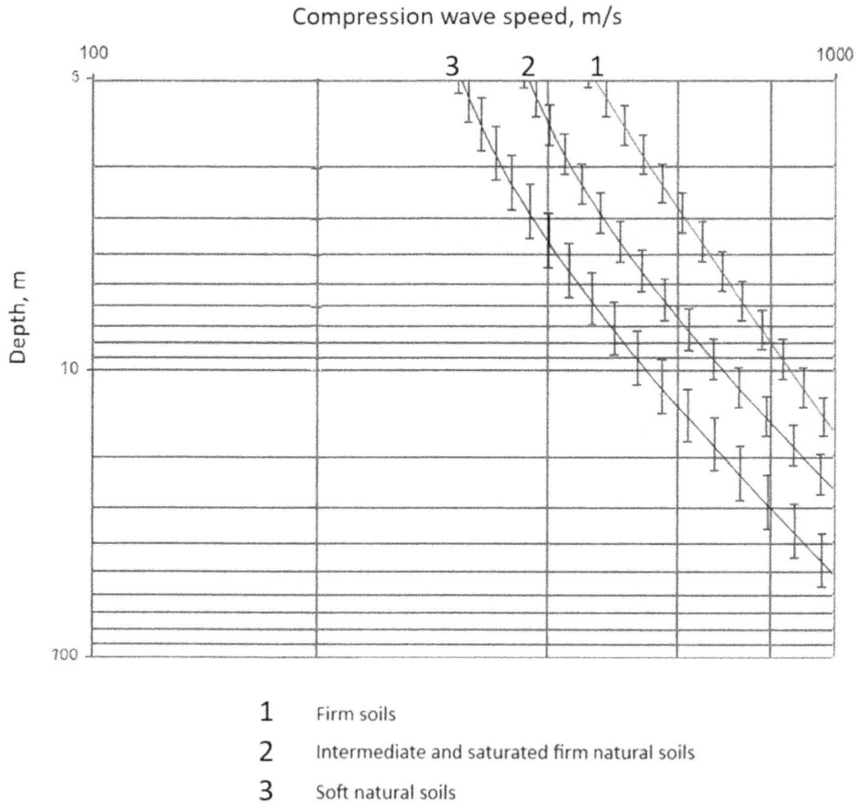

Figure 6.2 Compression wave speeds of soils as a function of depth.

floor which is $C_s = 78.98D^{0.312}$. This relationship is plotted in Figure 6.3. Hamilton gives regression equations for C_s as a function of four ranges of values of C_p which enable Poisson's ratio to be determined as in Table 6.5.

These values for Poisson's ratio, σ, are in three cases higher than the figure around 0.3 normally found, due to the effect of water saturation. Water raises the compression modulus by filling voids otherwise filled with air but has no effect on the shear modulus.

Close to the ocean floor, that is, at depths tending to zero, shear moduli cease to be directly relevant to shear moduli near ground level because water saturation normally begins some distance below ground. Ocean floor data can be considered relevant for present purposes at depths where the confining pressures and water saturation are similar to those under consideration.

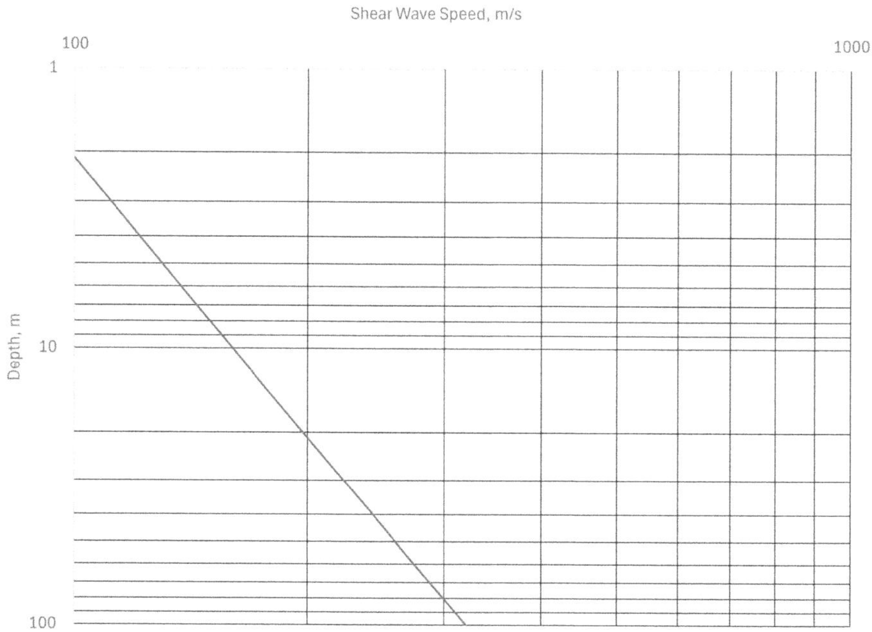

Figure 6.3 Shear wave velocity versus depth on water-saturated silt-clays.

Table 6.5 Poisson's ratios

V_p	V_s	σ
1555	282.62	0.211
1650	391.05	0.470
2150	721.175	0.437
3000	1378	0.366

In another paper[4] Hamilton gives a table of compressional wave velocity V_p, shear wave velocity V_p and Poisson's ratio σ in situ for water-saturated clays, turbidites and mudstones, sands, limestone and basalts. In the regression equations given, the depth is in km and the speeds in km/s (Tables 6.6–6.10).

Jafari[5] gives a summary of published relationships between shear wave speed C_s and SPT N value (Table 6.10). These are plotted in Figure 6.4.

Table 6.6 Wave speeds in water-saturated clays, turbidites and mudstones

Depth (m)	V_p, *m/s*	V_s, *m/s*	V_p/V_s	σ
0	1511	116	13.03	0.497
1	1512	121	12.50	0.497
5	1518	139	10.92	0.496
10	1524	163	9.35	0.494
20	1537	209	7.35	0.491
30	1549	256	6.05	0.486
40	1562	288	5.42	0.482
50	1574	301	5.23	0.481
100	1634	365	4.48	0.474
200	1744	438	3.98	0.466
300	1842	496	3.71	0.461
400	1930	554	3.48	0.455
500	2010	612	3.28	0.449
600	2082	670	3.11	0.442
700	2149	728	2.95	0.435
800	2212	786	2.81	0.428
900	2272	844	2.69	0.420
1000	2331	902	2.58	0.412

$V_p = 1.511 + 1.304D - 0.741D^2 + 0.257D^3$; $V_s = 116 + 4.65D(0 - 36\text{m})$;
$V_s = 237 + 1.28D(36 - 120\text{m})$; $V_s = 322 + 0.58D(120 - 650\text{m})$.

Table 6.7 Wave speeds in water-saturated sands

Depth (m)	V_p, *m/s*	V_s, *m/s*	V_p/V_s	σ
0.05	1727	55	31.40	0.4995
1	1806	128	14.11	0.4975
2	1825	155	11.77	0.4964
3	1836	174	10.55	0.4955
4	1844	189	9.76	0.4947
5	1850	201	9.20	0.4940
6	1855	211	8.79	0.4934
8	1863	229	8.14	0.4923
10	1869	244	7.66	0.4913
12	1875	257	7.30	0.4904
14	1879	268	7.01	0.4896
16	1883	278	6.77	0.4888
18	1886	288	6.55	0.4880
20	1889	296	6.38	0.4874

$V_p = 1806D^{0.015}$; $V_s = 128D^{0.26}$.

Table 6.8 Wave speeds in water-saturated limestone (measured)

Material	V_p, *m/s*	V_s, *m/s*	V_p/V_s	σ	*Pressure, kPa*
Calcite	6566	3414	1.92	0.31	10
Solenhofen	5560	2900	1.92	0.31	0
Limestone	6100	3200	1.91	0.31	0
	6300	3200	1.97	0.33	0
	5640	3000	1.88	0.30	10
	5370	2880	1.87	0.30	0
	5820	3070	1.90	0.31	0
	6100	3180	1.92	0.31	0
Bedford limestone	4680	1760	2.66	0.42	10
Spergen limestone	4700	2490	1.89	0.30	0
Carbonate-1	5961	3301	1.71	0.28	92
Carbonate-2	5840	3414	1.71	0.24	92
Carbonate-3	6261	3237	1.93	0.32	92
Carbonate-4	5936	3128	1.90	0.31	92
Carbonate-5	5609	2925	1.92	0.31	92
Carbonate-6	5589	2963	1.89	0.30	95
Carbonate-7	5491	2918	1.88	0.30	95
Limestone	4390	2570	1.71	0.24	20
	3730	2226	1.68	0.22	20
	4280	2145	1.99	0.33	20
	4995	2726	1.83	0.29	20
	5005	2812	1.78	0.27	20
	2813	1362	2.06	0.35	20
Leuders limestone	4119	2260	1.82	0.28	10

Table 6.9 Wave speeds in water-saturated basalts

Density (kg/m³)	V_p, *m/s*	V_s, *m/s*	V_p/V_s	σ
2100	3527	1732	2.04	0.34
2200	3747	1834	2.04	0.34
2300	3996	1955	2.04	0.34
2400	4274	2098	2.04	0.34
2500	4584	2266	2.02	0.34
2600	4929	2462	2.00	0.33
2700	5311	2690	1.97	0.33
2800	5731	2952	1.94	0.32
2900	6194	3253	1.90	0.31
3000	6700	3597	1.86	0.30

$V_p = 2.33 + 0.081(\rho/1000)^{3.63}$; $V_s = 1.33 + 0.011(\rho/1000)^{4.85}$.

Table 6.10 Relationship between shear wave speed and SPT N value

Author(s)	*Soil type*	C_s *(m/s)*
Kansai et al. (1966)	All	$19\ N^{0.6}$
Shibita (1970)	Sand	$32\ N^{0.5}$
Ohba & Toriuma (1970)	Alluvial	$85\ N^{0.31}$
Ohta et al. (1972)	Sand	$87\ N^{0.36}$
Ohsaki & Iwasaki (1973)	All	$82\ N^{0.39}$
Ohsaki & Iwasaki (1973)	Cohesionless	$59\ N^{0.47}$
Imai & Yoshimura	All	$92\ N^{0.329}$
Imai et al. (1975)	All	$90\ N^{0.341}$
Imai (1977)	All	$91\ N^{0.337}$
Ohta & Goto (1978)	All	$85\ N^{0.348}$
Ohta & Goto (1978)	Sands	$88\ N^{0.34}$
Ohta & Goto (1978)	Gravels	$94\ N^{0.34}$
JRA (1980)	Clays	$100\ N^{1/3}$
JRA (1980)	Sands	$80\ N^{1/3}$
Seed & Idriss (1981)	All	$61\ N^{0.5}$
Imai & Tonouchi (1982)	All	$97\ N^{0.314}$
Seed et al. (1983)	Sands	$56\ N^{0.5}$
Sykora and Stokoe (1983)	Granular	$100\ N^{0.29}$
Okamata et al. (1989)	Diluvial Sands	$125\ N^{0.3}$
Lee (1990)	Sands	$57\ N^{0.49}$
Lee (1990)	Clays	$114\ N^{0.31}$
Lee (1990)	Silts	$106\ N^{0.32}$
Imai & Yoshimura (1990)	All	$76\ N^{0.33}$
Yokota et al. (1991)	All	$121\ N^{0.27}$
Jafari et al. (1997)	All	$22\ N^{0.85}$

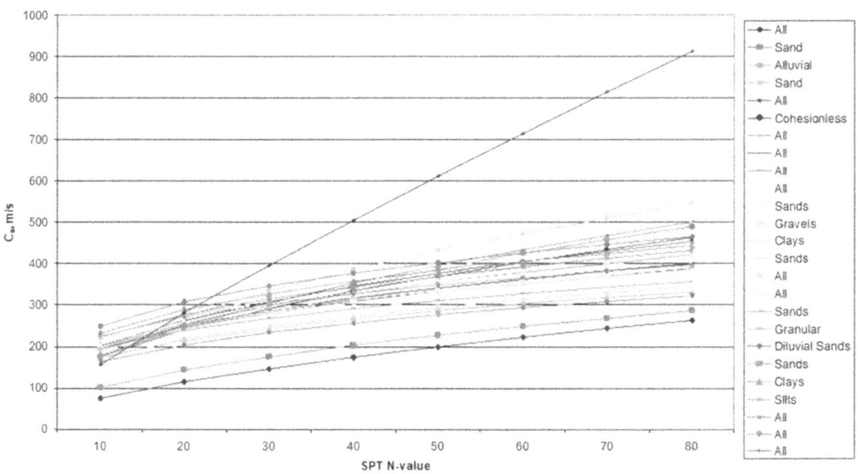

Figure 6.4 Damping as a relationship between SPT N value and shear wave speed.

NON-LINEAR SOIL MODULI

The non-linear behaviour of soil moduli was first expounded by Masing[6]; Madshus,[7] with reference to Isihara,[8] gives the Masing rule in Equations (6.2) and (6.3)

In the loading half cycle:

$$\frac{T_a + T}{2} = T\left(\frac{\Gamma_a + \Gamma}{2}\right) \tag{6.2}$$

In the unloading half cycle:

$$\frac{T_a - T}{2} = T\left(\frac{\Gamma_a - \Gamma}{2}\right) \tag{6.3}$$

where the brackets signify that T is a function of Γ as in Equations (6.4) and (6.5) according to the hyperbolic soil model of Hardin and Drnevich[9]

$$T(\Gamma) = G_{\text{max}} \cdot \frac{\Gamma}{1 + \frac{\Gamma}{\Gamma_r}} \tag{6.4}$$

G_{max} is the shear modulus for infinitesimally small strains; Γ is strain, and

$$\Gamma_r = \frac{T_f}{G_{\text{max}}} \tag{6.5}$$

where T_f is the shear strength

The stress-strain relationship thus defined is shown in the following example (Figure 6.5).

Hardin and Drnevich give an expression for G_{max} (Mpa) in terms of over consolidation ratio (OCR), void ratio, e, and effective mean principal stress σ'_m and plasticity index PI as in Equation (6.6):

$$G_{max} = 0.706266 \times \frac{(2.973 - e)^2}{1 + e} (OCR)^a {\sigma'_m}^{1/2} \tag{6.6}$$

where a is a function of plasticity index PI as in Table 6.11.

Hardin and Drnevich also presented relationships for the damping ratio which is principally for large strains.

Seed et al.[10] conclude that for most practical purposes, the dynamic shear moduli of granular soils (sands and gravels) can conveniently be expressed by the relationship

$$G = 6917 \cdot \mathrm{K}_2 \cdot (\sigma'_m)^{1/2} \tag{6.7}$$

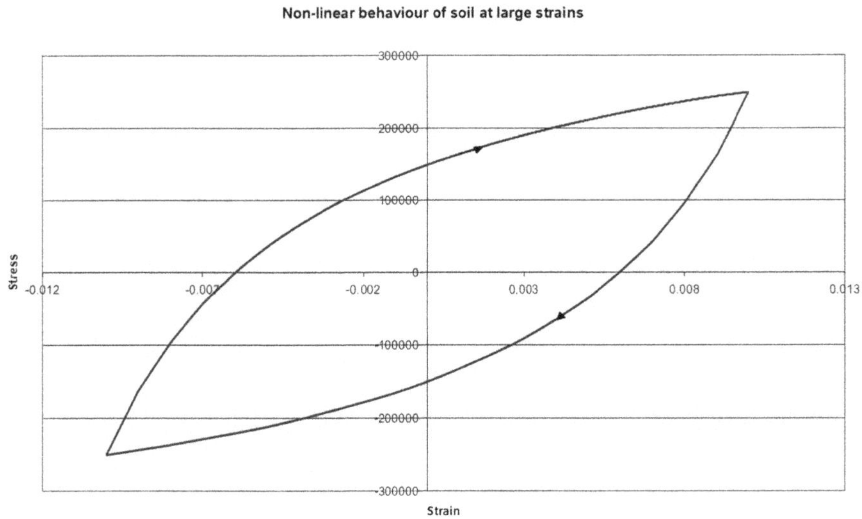

Figure 6.5 Non-linear behaviour of soil at large strains.

Table 6.11

PI	*a*
0	0
20	0.18
40	0.30
60	0.41
80	0.48
≥100	0.50

where K_2 is a shear modulus coefficient that is mainly a function of the grain size of the soil particles, the relative density of the soil, and the shear strain developed by the soil. The term σ'_m is the effective mean principal stress in Pascals (a function of depth). They give values of K_2 for a range of soils, and provide a guide to the determination of approximate values of K_2 in terms of the N-value measured in Standard Penetration Tests (SPTs) delivering 60% of the theoretical free-fall energy to the drill rods

$$(K_2)_{\max} \approx 30(N_1)_{60}{}^{1/3} \tag{6.8}$$

since SPT N-values are more readily available from geotechnical reports than other dynamic soil properties, this is a useful relationship.

Seed et al. give values of K_2 as in Table 6.12:

Table 6.12

Soil	*Location*	*Depth(m)*	K_2
Loose moist sand	Minnesota	3	34
Dense dry sand	Washington	3	44
Dense saturated sand	So. California	15	58
Dense saturated sand	Georgia	60	60
Dense saturated silty sand	Georgia	18	65
Dense saturated sand	So. California	90	72
Extremely dense silty sand	So. California	38	86
Dense dry sand (slightly cemented)	Washington	20	166
Moist clayey sand	Georgia	9	119
Sand, gravel, and cobbles with little clay	Caracas	60	90
Dense sand and gravel	Washington	46	122
Sand, gravel and cobbles with little clay	Caracas	78	123
Dense sand and sandy gravel	So. California	53	188

Seed and Idriss,[11] in an earlier paper than Seed et al., also give shear modulus data for peat in terms of the ratio of G to S_u, the undrained shear strength of 150.

DAMPING

The formula for the damping that results from the non-linear stress-strain relationship is

$$D_H = \frac{4}{\pi}\left(1 + \frac{\Gamma_r}{\Gamma_a}\right)\left(1 - \frac{\Gamma_r}{\Gamma_a} \cdot \ln\left(1 + \frac{\Gamma_a}{\Gamma_r}\right)\right) - \frac{2}{\pi} \tag{6.9}$$

This relationship is plotted in Figure 6.6. for a range of ratios of $G_{\max}$ to undrained shear strength. It can be seen that the damping due to Masing behaviour at small strains is very small.

Hamilton plots a range of data on attenuation versus frequency for natural, saturated sediments and sedimentary strata. The upper and lower bounds are 0.075 to 0.003 dB per metre at 100 Hz, increasing in direct proportion to frequency. As already mentioned, a loss factor that is independent of frequency gives attenuation in dB per metre which is proportional to frequency. Stoll[12] found the loss factor of shear waves in a water-saturated silt to be independent of frequency between 43 and 391 Hz. Tullos and Reid[13] found that in clay-sand the loss factor is independent of frequency between 20 and 400 Hz. In sands, Stoll found that loss factor was proportional to frequency (and, therefore, attenuation proportional to the square of the frequency) in the range 30–240 Hz. This suggests the existence of viscous damping.

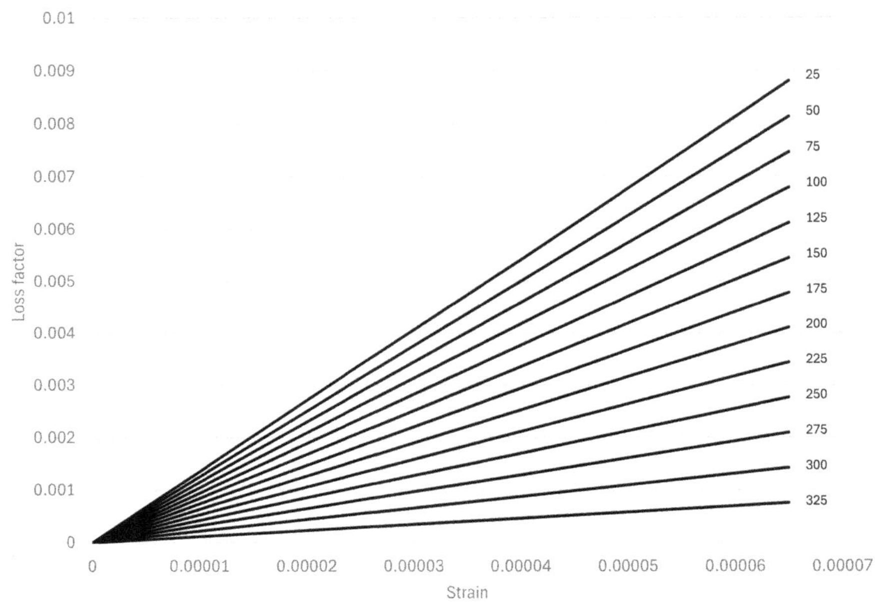

Figure 6.6 Damping as a function of strain.

Hamilton gives data on attenuation in dB per metre as a function of porosity in saturated surface (of the ocean floor) sediments. There is a peak attenuation at 52% porosity of 0.78 f dB per metre, where f is frequency in kilohertz, falling to 0.12 f at 65%, 0.52 f at 46%; 0.46 f at 36% and 005 f at 90%. Similarly he gives attenuation as a function of mean grain size peaking at 0.76 f at 40μ.

Hamilton quotes attenuation in hard, dense limestone of 0.02 f and chalk 0.08 f, dB per metre. For Basalts under the sea floor the attenuation is given as 0.02 f to 0.05 f. Hamilton summarises various references as in Table 6.13:

Table 6.13

Material	*Attenuation in dB per metre (f = kHz)*
Diluvial sand	13.2 f
Diluvial sand and clay	4.8 f
Alluvial silt	13.4 f
Mud (silt-clay)	17.3 f
Water-saturated clay	15.2 f
Tertiary mudstone	10.1 f
Pierre shale	3.4 f
Solenhofen limestone	0.02 to 0.05 f
Chalk	0.1 f
Basalt	0.07 f

Stoll[14] shows loss factor for dry Ottawa sand and saturated Ottawa sand which are reproduced in Figures 6.7 and 6.8.

The topic of sound propagation in fluid-saturated saturated porous solids was pioneered by Biot[15,16] and Stoll[14] and is discussed in detail in Chapter 3.

The Biot-Stoll formula gives a compressional wave that is always equal to, or higher than, the conventional elastic compression wave, and a second compressional wave which is slow and highly attenuated. While the second kind of wave can be ignored as a received signal, it is important in that a fractional of the initial wave energy is carried by it, and rapidly attenuated, reducing the amplitude of the fast wave. In addition, on reflection at boundaries between soils of different impedances, some of the compression wave energy will be reflected as a wave of the slow kind, also to be rapidly attenuated. This may explain why field measurements of attenuation with distance from ground level or underground sources suggest higher loss factors than can be accounted for by friction, viscous or Masing effects.

By comparison, the Biot-Stoll formula for shear waves is more simply related to the classical shear wave speed for elastic media.

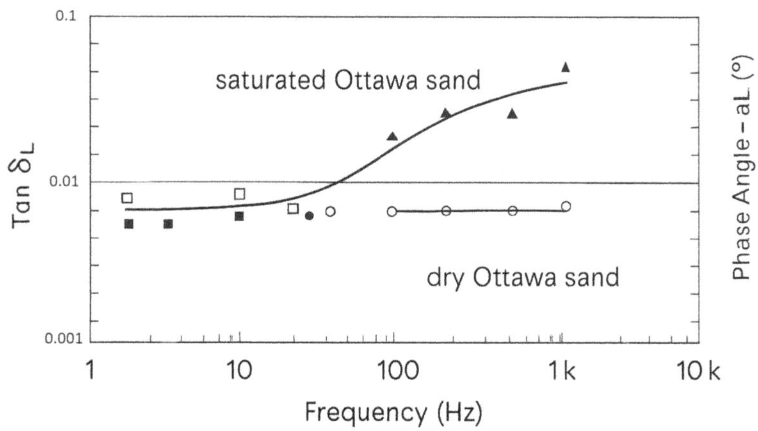

Figure 6.7 Loss angle from experiments on 20–30 Ottawa sand.

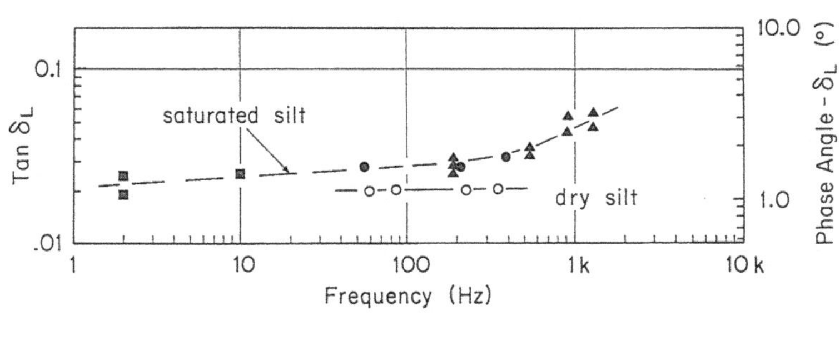

Figure 6.8 Loss angle from experiments on inorganic silt.

Stoll noted that, when strain amplitudes are less than about 0.0001% soil, moduli can be regarded as linear, and simpler equations for predicting velocity and attenuation are possible. He noted that in the case of dry materials such as sands, silts, sandstones and other granular materials, attenuation in low-amplitude waves may be adequately described by a model that assumes a constant loss factor. The principal mechanism of energy loss in these dry materials is friction, which governs the minute amount of slip occurring within the contact area between particles. As a result the damping observed in a granular material is usually very small (e.g., typically of the order of 0.01 in sandstones and sands). When the pore spaces between the particles contain a fluid such as water, the observed attenuation is often significantly larger, due to the addition of viscous losses in the fluid. In the case of materials with high permeability, there is overall motion of the fluid field relative to the skeletal frame. This result is frequency-dependent damping which changes rather rapidly over a fairly narrow frequency range. In finer materials with very low permeability, this kind of damping does not appear until quite high frequencies are reached. Another kind of viscous damping occurs, however, even in the low frequency range, as a result of local fluid motion in the vicinity of the intergranular contacts similar in nature to "squeeze film motion", as in the theory of lubrication. This second kind of viscous damping causes frequency-dependent damping that can be described in terms of classical viscoelastic models. Stoll's summary of measured and predicted data for compressional waves is given in Figure 6.9.

Seed et al. give damping data for sands and gravels, but mainly for strains greater than 0.001%.

Darendeli, Stokoe and Gilbert[17] give an expression for damping at infinitesimally small strains, D_{min}, as in Equation (6.10):

$$D_{min} = \left(\phi_6 + \phi_7 \times PI \times OCR^{\phi 8}\right) \times {\sigma'_o}^{\phi 9} \times [1 + \phi_{10} \times \ln(f)] \qquad (6.10)$$

where PI = Plasticity Index (%), OCR = over consolidation ratio σ'_o = mean effective confining pressure (atm) and f = loading frequency, $\phi_6 = 0.8005$; $\phi_7 = 0.0129$; $\phi_8 = -0.1069$. $\phi_9 = -0.2889$ and $\phi_{10} = 0.2919$.

This gives values for D_{min} at 1 Hz of between 1.8% and 2.1% (loss factors of 0.036 to 0.041) for PI from 16 to 1, and a plasticity index of 100%, falling linearly to zero at PI = 0. D_{min} increases by a factor of 2.3 if the frequency is increased from 1 Hz to 100 Hz.

Shear modulus can be determined both from data on shear modulus itself, and from data on shear wave speed, but there are few data on gravel.

For non-marine conditions, Lew and Campbell give shear and compression moduli for 10–50% gravel at around 400 m/s and 750 m/s respectively, which, for a density of 1600 kg/m^3 leads to $G = 0.256$ GPa and $D = 0.9$ GPa. Green gives attenuation coefficients of 0.03 to 0.09 dB per metre at 5 Hz and 0.26 to 0.87 dB/metre at 50 Hz. For the wave speeds of Lew and Campbell, these equate to loss factors of 0.09 to 0.26.

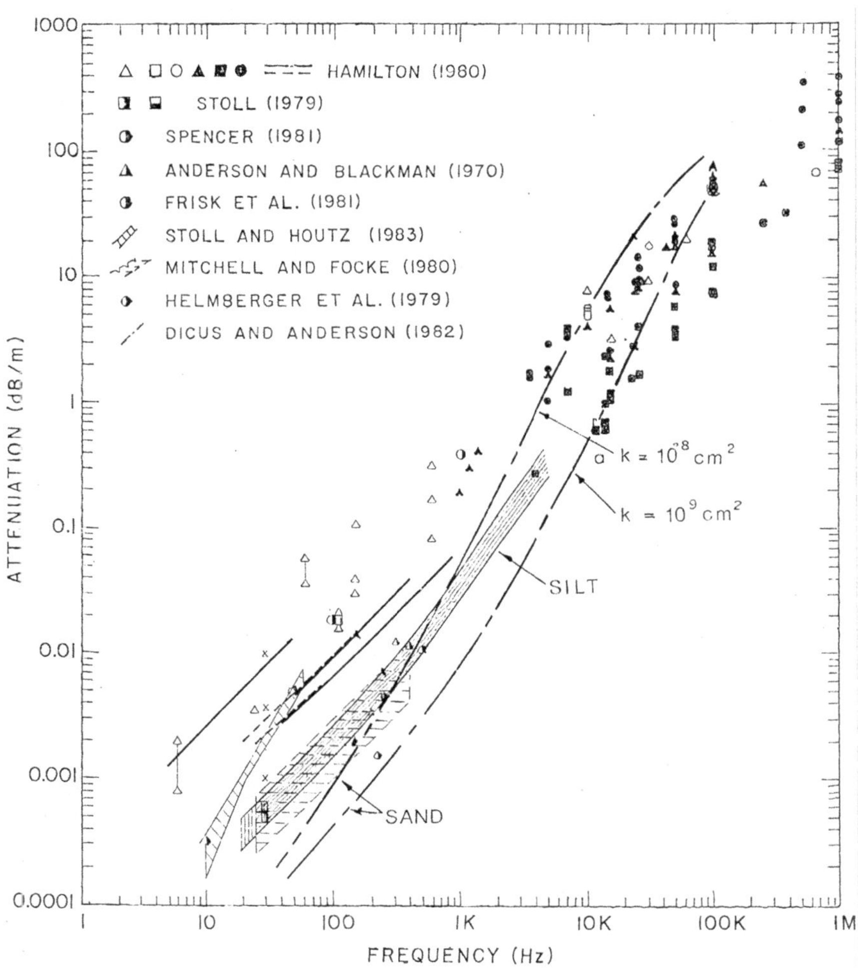

Figure 6.9 Laboratory and field data for attenuation versus frequency with curves showing predictions of Biot model for sand and silt superimposed.

Seed's method of estimating G_{max} gives a figure for "Sand, gravel and cobbles with little clay" at a depth of 60 m of 0.48 GPa. This gives a surface figure of about 0.2 GPa. Ohta and Goto give a relationship between C_s and SPT value for gravel. Assuming a SPT N value of about 30, this gives a value for C_s of 300 m/s. This gives $G = 0.144$ GPa.

Lew and Campbell give a value of C_p for unspecified "saturated soil" of 1605 m/s, which leads to a value for D of 4.1 GPa and a Poisson's ratio of 0.47. The Biot-Stoll formulae give values somewhat higher than this for clays and sands.

Theory suggests that loss factors at low frequencies are unlikely to be as high as 0.1 and that a figure of 0.05 is the most that can be justified.

On the other hand, some additional excess attenuation will occur due to wave mode conversion at reflective boundaries between different soil types.

The characteristics of made ground obviously depend on its nature. A conservative approach would be to assign to it the same properties as those for gravel.

NOTES

1. Jardine, R.J., Potts, D.M., Fourie, A.B., and Burland, J.B. "Studies of the influence of noon-linear stress-strain characteristics in soil-structure interaction." *Géotechnique* 36(3), 377–396.
2. Lew, M., and Campbell, K.W. "Relationship between shear wave velocity and depth of overburden." In *Measurement and use of shear wave velocity for evaluating dynamic soil properties*. Woods, R.D. (ed.), Proceedings of a Session Sponsored by the Geotechnical Engineering Division of the American Society of Civil Engineers in conjunction with the ASCE Convention in Denver, CO, 1985.
3. Hamilton, E.L. "Geoacoustic modelling of the sea floor." *The Journal of the Acoustical Society of America*, 68(5), 1313–1340, November 1980.
4. Hamilton, E.L. "V_p/V_s and Poisson's ratios in marine sediments and rocks." *The Journal of the Acoustical Society of America*, 66(4), 1093–1101, October 1969.
5. Jafari, M.K., Shafiee, A., and Razmkhah, A. "Dynamic properties of fine grained soils in South of Tehran." *Journal of Seismology and Earthquake Engineering*, 4(1), 25, Spring 2002.
6. Masing, G. "Eigenspannungen und Verfestigung beim Messing." *Proceedings, Second International Congress of Applied Mechanics*, Zurich, 332–335, 1928.
7. Madshus, C. "Soil non-linearity and its effect on the dynamic behaviour of offshore platform foundations." D.Sc. Dissertation, Department of Geology, Faculty of Mathematics and Natural Sciences, University of Oslo, 1997.
8. Isihara. K. "Evaluation of soil properties for use in earthquake response analysis." *Proceedings: International Symposium on Numerical Models in Geomechanics*, Zurich, 13-17 September 1982.
9. Hardin. B.O., and Drnevich, V.P. "Shear modulus and damping of soils, design equations and curves." *Journal of the Soil mechanics and Foundations Division*, 98(SM7), 667–692, 1972.
10. Seed, H.B., Wong, R.T, Idriss, I.M., and Tokimatsu, K. "Moduli and damping factors for dynamic analyses of cohesionless soils." *Journal of Geotechnical Engineering*, 112(11), 1016–1032, 1986.
11. Seed, H.B., and Idriss. I.M. "Soil Moduli and damping factors for dynamic response analyses." Earthquake Engineering Research Centre, Report No EERC 70-10, December 1970.
12. Stoll, R.D. "Experimental studies of attenuation in sediments." *The Journal of the Acoustical Society of America*, 64, 5143(A), 1978.
13. Tullos, F.N., and Reid, C. "Seismic attenuation of gulf coast sediments." *Geophysics*, 34, 516–528, 1969.
14. Stoll, R.D. *Computer-aided studies of complex soil moduli*. Woods, R.D. (ed.), Proceedings of a Session Sponsored by the Geotechnical Engineering Division of the American Society of Civil Engineers in conjunction with the ASCE Convention in Denver, CO, 1985.

15. Biot, M.A. "Theory of propagation of elastic waves in a fluid-saturated porous solid – I low-frequency range." *The Journal of the Acoustical Society of America*, 28(2), 168–178, 1956.
16. Biot, M.A. "Theory of propagation of elastic waves in a fluid-saturated porous solid – II higher frequency range." *The Journal of the Acoustical Society of America*, 28(2), 179–191, 1956.
17. Darendeli, M.B., Stokoe, K.H., and Gilbert, R.H., *Development of a new family of normalized modulus reduction and material damping curves*. University of Texas, Austin, TX, 2001.

Chapter 7

Power flow

Vibration is normally considered because of its potential effect on human beings and/or sensitive equipment. Effects caused by vibration are assessed using scales that measure motion, namely acceleration, velocity or displacement, and transducers for measuring vibration normally measure either acceleration or velocity. However, in the process of predicting vibration it is necessary to calculate transmission functions between source and receiver, and these must take into account energy transmission, reflection and absorption in the course of calculating motion-based amplitudes at the receptor. It is frequently the case that, at a far point in a transmission line, the velocity amplitude is greater than at a nearer point, because the impedance of the local medium is lower (i.e., the mobility is higher) at the far location than at the near location. Even if the power flow in the far location is lower, the velocity associated with that power may be significantly greater than in the nearer location. A commonly encountered case is the rise in velocity amplitude when moving from a ground-bearing slab to a suspended first floor. The detailed application of power flow concepts to vibration is a developing field, and the availability of more powerful computation techniques make it possible to consider more easily vibration generation, transmission and reception in power terms. This chapter will review the underlying principles, and present the results of some power flow analyses of vibration problems.

In the discussion of the propagation of groundborne noise in Chapter 3, vibration was largely quantified in terms of dimensional amplitude, that is, displacement, velocity and acceleration. Indeed, vibration if this kind is normally measured in such terms, and when perceived by a human being, or received by an item of vibration-sensitive equipment, the magnitude of perception or effect is often proportional to acceleration, velocity or displacement, albeit subject to frequency weighting.

However, in order to understand the underlying behaviour of a complete system, from source through a path to a complex receiver, it is necessary to consider not just dimensional amplitudes, but also power flow. When vibration enters a building, it will give rise to vibration velocity in surfaces close to

DOI: 10.1201/9781003618775-7

the point of entry which may be exceeded significantly in other parts of the building, due to what is loosely referred to as "amplification" in lightweight parts of the structure, such as suspended floors. Actually no amplification, in the sense of the word when used in audio systems, occurs. No extra energy is introduced. What happens is that the mobility (the inverse of impedance) of a lightweight structure is greater than the mobility of a concrete foundation or slab and for a given value of the power in the signal, the velocity (and other dimensional measures) will be higher.

In order to understand this further, some fundamental principles must be considered:

- Power is rate of doing work
- Work is force multiplied by distance (in units of Nm or Joules)
- Power is Nm/s or joules per second or watts
- Impedance is force/velocity
- Mobility is velocity/force
- Power $= \overline{\text{force} \times \text{velocity}} = \overline{\text{velocity}^2} \times \text{impedance}$
- Power $= \overline{\text{velocity}^2}/\text{mobility}$

where the bar above the words "force" and "velocity" indicates the mean value).

For those familiar with electrical engineering, the relationships are analogous to "volts times amps equals watts" or $VI = W$ and "volts equals current times impedance" or $V = IZ$. Voltage is the analogy of force or pressure, and current is the analogy of velocity. Impedance in electrical engineering may be inductance, capacitance or resistance, and these are analogous to mass, spring rate and damping. Phase, in electrical engineering, has the same meaning as in vibration. The relationship $VI = W$ is straightforward in the case of direct current or DC. In vibration, the analogy is to alternating current, AC, and both voltage and current are expressed in terms of both amplitude and phase, using complex notation. The complex power S is VI^* where V is the complex voltage and I^* is the complex conjugate of the current. The complex conjugate has the sign of the imaginary part reversed. Complex power consists of resistive (real) and reactive (imaginary) parts, and only the resistive part relates to energy transmitted through the circuit. The reactive part relates to energy periodically stored and restored, and not transmitted.

When the quantity of interest is power flow, $\overline{P} = \frac{1}{2}|\tilde{v}|^2 R$ where R is the real part of the impedance, that is, the resistance-like part, and $|\tilde{v}|^2$ is the mean-square velocity.

Impedance becomes important in cases where it is necessary to consider changes to a system, for example, when resiliently mounting a vibration-sensitive device on a floor that is set in vibration by an external source.

The vibration of the floor can be measured in the absence of the mounted device, and it is necessary to calculate the vibration of the device after it has been installed on its mountings. The installation of the device will change the vibration of the floor. The reaction force on the floor resulting from the presence of the device on its mountings is $\tilde{Z}_1\tilde{v}_D$ where $\tilde{v}_D$ is the floor velocity in the presence of the mounted device and $\tilde{Z}_1$ is the impedance of the mountings of the device and the device itself. Thus, the relationship between the floor velocity in the absence of the device, $\tilde{v}_F$, and the floor velocity with the mounted device present, is given in terms of the floor impedance $\tilde{Z}_F$ as

$$\tilde{v}_D = \tilde{v}_F\left[1 + \tilde{Z}_1/\tilde{Z}_F\right]^{-1} \tag{7.1}$$

If the device is simply a mass on a spring, this equation shows its behaviour as a dynamic absorber.

Electrical analogies can be both helpful and a source of confusion. When considering impedances, the analogies to mass-like elements and spring-like elements are inductance and capacitance, the impedances of which are imaginary numbers. The impedance of a mass is $i\omega m$ and the impedance of a spring is $k/i\omega$. The analogy of a damper is resistance, the impedance of which is a real number. When considering matters such as power flow, it is the real part of an impedance that determines the power since the imaginary part represents elements that store energy but return it so that there is no net contribution to the power flow. This makes sense in a system such as a mechanical structure. However, when considering power flow in free-field, such as the power transmitted by a plane wave travelling through an infinite isotropic medium, such as rock, the characteristic impedance of the rock, if it is for the sake of argument regarded as having a zero loss factor, is a real number given by ρc. As a result, force and velocity in the loss-free rock are in phase, and power is transmitted without attenuation due to distance over and above that due to geometric spreading. Yet, normally, a real impedance is associated with resistance, which, by definition, involves energy loss, that is, absorption. The question arises: Why does a real impedance allow loss-free transmission? The answer lies in the parameters involved in deriving the impedance of the medium. The impedance ρc can also be written as $\sqrt{(\rho D)}$ where ρ is the mass density and D is the compression modulus. The impedance of the mass-like parameter is $i\omega\rho$ and the impedance of the spring-like parameter D is $D/i\omega$. It is when these two imaginary impedances are multiplied together (and the square root taken) we get a real frequency-independent number $Z = \sqrt{(i\omega\rho D/i\omega)}$, because the two $i\omega$ terms cancel. It is for the same reason that an infinite loss-free electricity transmission line has a real-numbered characteristic impedance, even though the impedance

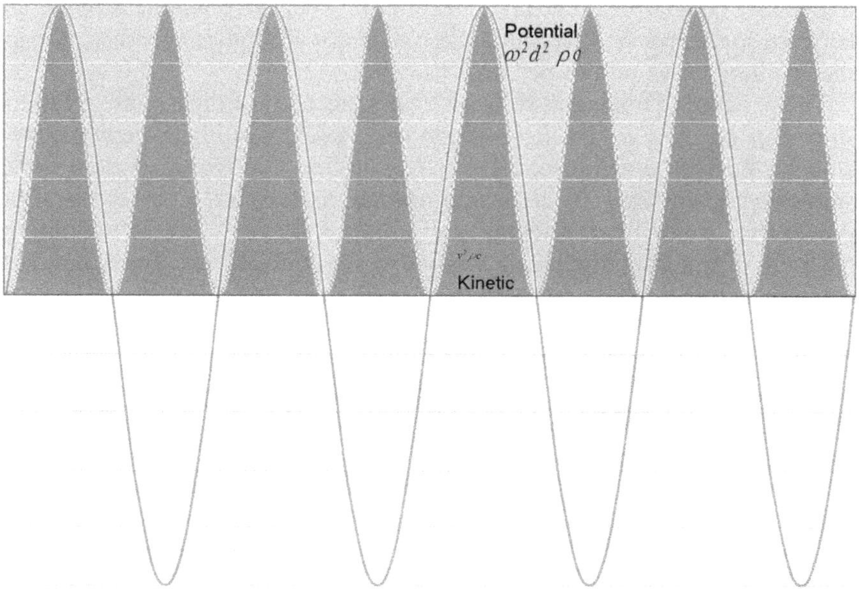

Figure 7.1 Kinetic and potential energy in a wave propagating through a medium with impedance ρc (v = velocity and d = displacement).

results from inductance of the cable and capacitance between the cable and the ground. Thus, the impedance of the rock is a real number, even though that number actually represents energy storage characteristics and no energy absorption characteristics. It is not always true to say that real numbers represent resistive characteristics, and imaginary numbers represent reactive characteristics. When a wave is propagated through this medium, energy is alternately stored as kinetic and potential energy, being swapped between the two twice in each cycle (in the case of a single frequency signal) (Figure 7.1).

Energy absorption only occurs if the parameter D becomes complex and, as explained in Chapter 3, is expressed as $D(1 + i\eta)$ where η is the loss factor. The square root of $(1 + i\eta)$ is for $\eta < 1 \approx 1 + i\eta/2$, so the impedance of the medium when loss is added becomes $\sqrt{\rho D}(1 + i\eta/2)$. A description of the two parts of a complex compression modulus or $D' + D''$ is that D' represents the storage modulus, and D'' the loss modulus. The effect of the loss modulus appears in the propagation equation e^{ikx} which then has two parts, an oscillating part and an attenuating part. When considering the power flow through the infinite isotropic medium just discussed, in the loss-free case the force and velocity are in phase, and when multiplied together to get power, the resulting curve is all positive. When a complex impedance is assumed, that takes account of loss, there will be a phase lag between force and velocity, and, when they are multiplied together, part of the resulting curve will be negative.

The product of force and velocity, taking account of their phase relationship, can be used in another, sometimes useful, way. In a network-type system, such as a reinforced concrete building consisting of floor slabs supported on columns, it can be important to understand whether vibration power from a nearby or underground source is flowing into the building in all parts of the structure, such as piles in the foundations. For a building extending over a large area, vibration power can flow into the building at the front nearest to the source, and out of it back into the ground at the rear. This can be important if consideration is being given to providing base-isolation for the building, and whether bearings are needed over all sides, or only parts nearest to the source. Leaving the further parts unisolated can be beneficial if they are acting as "sinks" and extracting vibrational energy from the system by letting it flow back into the ground. If a coordinate system is used in which the x-direction of velocity is into the building, then the positive parts of the $v \times f$ curve will represent inward power flow, and negative parts outwards.

Measurement of both force and velocity simultaneously is not an easy matter. The usefulness of, for example, methods of determining power flow direction is primarily in the application of numerical modelling methods where both properties can be output from a model. Measurement in-situ is made difficult by the fact that while velocity can be readily derived from measurement of acceleration (using accelerometers) or velocity (using geophones), measurement of force requires the use of transducers such as strain gauges or force transducers. In some cases, the use of two or more accelerometers can be used – for example, to study the behaviour of a resilient bearing supporting a building. One accelerometer can be used to determine velocity, while the other(s) can be used, after double-integrating the signal or dividing the Fourier spectrum by ω^2, to determine the displacement of the top and bottom of the bearing, so that, from knowledge of the complex frequency-dependent, stiffness of the bearing, the force can be derived.

The fact that the impedance of the infinite medium is a real number becomes interesting if the case changes to a semi-infinite half-space with a structure supported by it. The equations for transmission ratio or factor given in Chapter 3 relating to dimensional amplitude such as velocity look very different when expressed as ratios of transmitted to incident power (Figure 7.2).

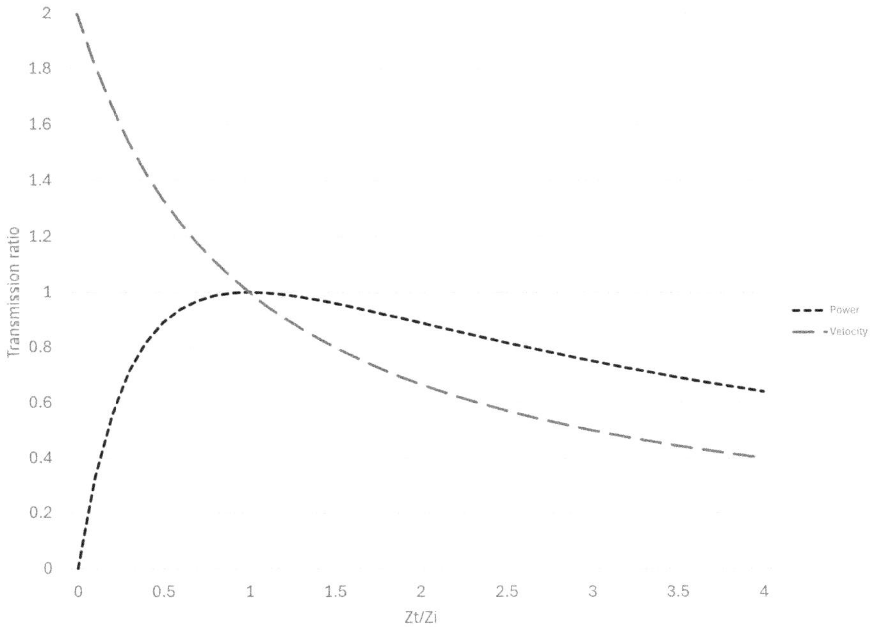

Figure 7.2 Relationship between velocity transmission and power transmission.

GROUP VELOCITY

In Chapter 3 the range of wave types that occur in viscoelastic media were considered and, in particular, their speed of propagation was calculated, being in the simple case of *p*-waves in an infinite homogenous isotropic medium as $\sqrt{D/\rho}$ where D is the compression modulus and ρ is the mass density. This wave speed is specifically known as the phase speed, meaning that if you look along a time sequence of plots of the waveform and pick the times and distances at which the phase of the wave is the same, after displacing it by the elapsed time and distance travelled, the distance divided by the time is the phase speed.

There is, however, another kind of wave speed that occurs when the principal wave is acting as an amplitude-modulated carrier wave. The simple case exists when two waves of slightly different frequencies and wavenumbers are superimposed, causing a "beating" effect. The "beat" has a velocity C equal to

$$C = \frac{\omega_1 - \omega_2}{k_1 - k_2} \tag{7.2}$$

where ω_1 and ω_2 are the angular frequencies of the first and second waves and k_1 and k_2 are the corresponding wave numbers. The group velocity can be considerably larger than, or considerably smaller than, the phase velocity.

The other important case is when the phase speed is frequency dependent, which happens in bending waves in plates and beams, and also with Rayleigh waves when the soil is not homogeneous and isotropic.

The group velocity equation can be written as

$$C = \frac{d\omega}{dk} \tag{7.3}$$

In thin beams, where the phase speed of bending waves is proportional to the square root of frequency, the group velocity is simply twice the bending wave velocity. In thin plates, at the frequencies normally encountered in groundborne noise the same is true, but, at higher frequencies, the group velocity converges towards the phase velocity.

The main consequence of the dispersion of wave velocities in beams and plates is that power flow is given by the product of the group velocity and the energy density of the wave packet.

Chapter 8

Conversion of groundborne vibration into audible sound

In the field of groundborne noise, it is frequently necessary to predict the relationship between the vibration of the surfaces of a room and the sound level in the room resulting from radiation from the room surfaces acting as "loudspeakers". In the simplest of cases, an infinitely large plane surface vibrating as a piston, i.e. all parts of its surface vibrating in phase so that it has a radiation ratio of 1 will result in a sound pressure level in air adjacent to it equal to $L_v - 20\ \log_{10}\left(\rho c\ V_{ref}/p_{ref}\right)$, that is, $L_v - 33.8$ dB where L_v is dB re 1 nanometre per second. If a hypothetical room floor had a radiation ratio of 1, a zero reverberation time and no other surface were radiating, $L_v - 33.8$ dB would be the relationship between floor vibration and sound level. In reality, other surfaces in the room will be contributing, albeit possibly to a smaller extent than the floor, and the reverberation time will be non-zero. The radiation ratio of the surfaces may be less than 1 at least at some frequencies. It is also necessary to take into account that reverberation time, as normally defined for the case of a diffuse field in a reverberant room whose dimensions are larger than the relevant wavelengths of the sound, will follow different rules, as modal acoustics theory is required when room dimensions and wavelengths are comparable.

Because of these complications, a practical approach has traditionally been taken, relying on field measurements reported in the literature, giving results for the vibration-to-sound correction. Kurzweil[1] reviewed the topic and reported that a reasonable estimate could be made of the relation between the octave-band room floor acceleration level, and the resulting octave band sound pressure level in the room. When expressed in terms of velocity level, Kurzweil's equation simplifies to

$$L_p = L_v - 27\ \text{dB} \tag{8.1}$$

where L_v = floor vertical velocity level in dB re 1 nanometre/second and L_p = sound pressure level in the room in dB re 20 μPa. This relationship has been widely used, including a similar version used in the USA that $L_p = L_v$ when L_v is in dB re 1 microinch/second, equivalent to $L_p = L_v - 28$ dB,

DOI: 10.1201/9781003618775-8

although Zapfe et al.[2] reported that a better model was $L_{pAwt} = L_{vAwt} - 33$ dB where the "Awt" subscript signifies the overall A-weighted level. However, they presented a group of measured results plotted against this equation that showed some measurements over 10 dB higher, with the highest being 15 dB higher. Kouroussis et al.[3] and Vogiatzis[4] proposed $L_p = L_v - 22$ dB with a worst-case assumption of $L_p = L_v - 16$ dB.

One of Kurzweil's main reference sources, the US Department of Transportation 1974 Report "Prediction and control of rail transit noise and vibration – a state-of-the-art assessment" reported that, applying a correction of −28 dB, there is fair agreement between measured and predicted results up to a frequency of 125 Hz, but a 12 dB underprediction at 250 Hz. Figure 8.1 comes from the report from the RIVAS[5] project and shows the amount by which noise level exceeds vibration level in terms of the mean, and the mean ± one standard deviation. This chart uses the 50 nanometre/second reference, so 34 dB should be subtracted. It shows that while the mean correction is about −30 dB ± 6dB at 63 Hz it is −21 dB ± 9 dB at 125 Hz. This means that at −12 dB 20% of results will be above (worse than) the prediction. Figure 8.3 comes from PD TS 14837-31:2017 "Guideline on field measurements for the evaluation of human exposure in buildings" and shows that the correction can be an average figure close to −22 at 125 Hz, but with a maximum of −17 dB.

A Transit Cooperative Research Program (TCRP) report[6] shows a very wide distribution of relationships between vibration level and sound level.

A more detailed approach is to calculate the relationship using wave acoustics, or to model the system numerically.

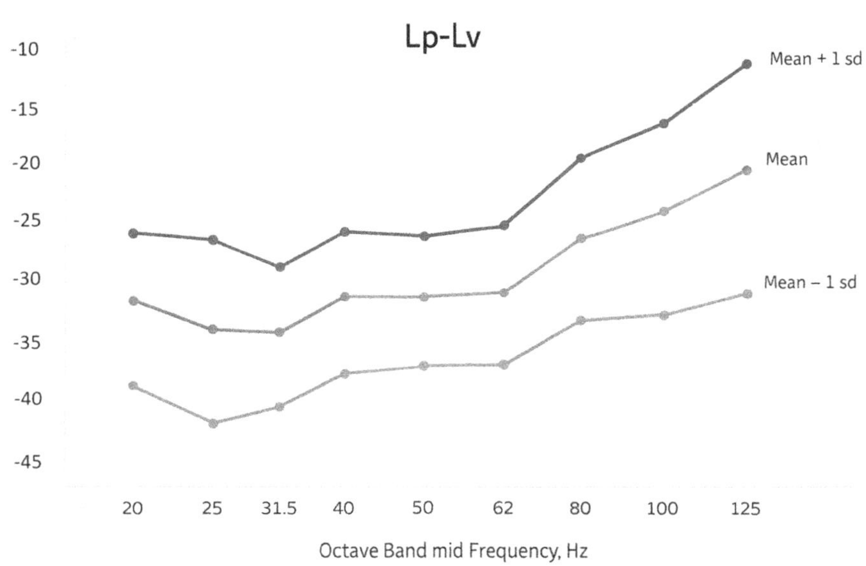

Figure 8.1 Sound pressure level minus velocity level based on RIVAS report in Reference 5.

SOUND RADIATED BY A PLATE INTO A RECTANGULAR ROOM

The sound pressure level that occurs in a room, one surface of which is a vibrating plate, is a function of the plate mode shape function and the room mode shape function. The coupling between them is determined by the product of the two functions integrated over the surface of the plate. Let us consider a room with dimensions L_x, L_y and L_z, where L_z is the height of the room and one surface in the x, y plane, that is, the floor, is vibrating.

For a simply supported plate the shape function is[7,8]

$$\psi(x,y) = \sin\frac{m\pi x}{L_x}\sin\frac{n\pi y}{L_y} \tag{8.2}$$

where m and n are any integer greater than 0. This is associated with plate eigenfrequencies which are given by

$$\omega_N = \sqrt{\frac{D}{\rho_s}}\left[\left(\frac{m\pi}{L_y}\right)^2 + \left(\frac{n\pi}{L_x}\right)^2\right] \tag{8.3}$$

where $D = \frac{Eh^3}{12(1-\sigma^2)}$ is the bending stiffness, ρ_s is the surface density, σ is Poisson's ratio, h is the plate thickness and E is Young's modulus.

Mode 1,1 is the lowest plate mode with an antinode in the centre of the plate, although mode 0,0 can exist if the plate is moving vertically as a piston, with consequences discussed below. If a floor plate is vibrating as a piston, it will have increased coupling with the air in the room, but only at room mode 0,0,r. In the case of clamped edges, the frequency of the lowest eigenmode will be much higher than in the case of simply supported edges, often above the frequency range in which groundborne noise from underground railways has its peak.

The room mode shape function is[9]

$$\phi(x,y,z) = \cos\frac{p\pi x}{L_x}\cos\frac{q\pi y}{L_y}\cos\frac{r\pi z}{L_z} \tag{8.4}$$

where p, q and r are any integer including 0.

This is associated with the room eigenfrequencies that, for a lossless room, are given by

$$\omega_N = \pi c\sqrt{\left(\frac{p}{L_x}\right)^2 + \left(\frac{q}{L_y}\right)^2 + \left(\frac{r}{L_z}\right)^2} \tag{8.5}$$

The amplitudes of the shape functions on a decibel scale calculated using Equations (8.2) and (8.4) are plotted in Figure 8.2 for modes 1,0 (room only) 1,1 2,1 2, 2 in the xy plane. Mode 0,0, not plotted, is of equal level across the plane.

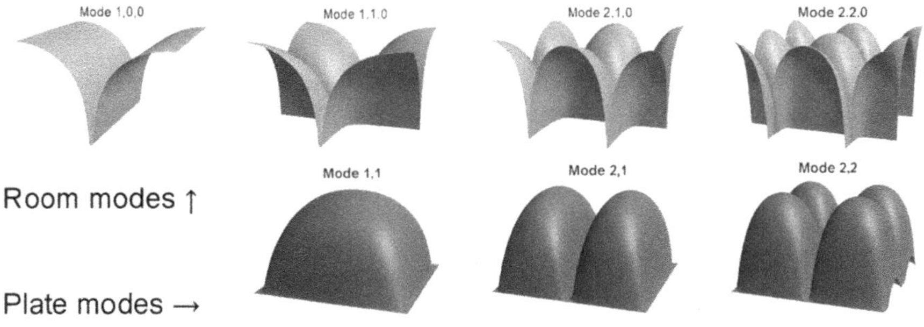

Figure 8.2 Room shape functions and corresponding plate mode shape functions (plotted as logarithmic squared amplitude).

Mode 0,0,0 occurs when the air in the room is being compressed, but the forcing frequency of the compression is too low, and therefore the wavelength too long, to cause a standing wave. The eigenfrequency for mode 0,0,0 is zero. Given that the floor plate has a finite mass, in reality mode 0,0,0 is the mass-air resonance of the floor plate as a mass on an air spring, that is, $\omega_{0,0,0} = \sqrt{\left(\frac{L_x L_y \rho c^2}{L_z \rho_s}\right)}$.

Equations (8.2) and (8.4) are widely used in airborne acoustics in considering coupling between walls and rooms although other functions are required for other edge conditions.[10]

Coupling between the plate and the room is given by integrating the product of the plate shape function and the room shape function for $z = 0$ where z is the room height if the plate is the floor. This integral leads to the result that no coupling occurs when $m + p$ or $n + q$ is even. This, however, does not take into account the fact that in the case of room heights that are less than the other room dimensions, which is usually the case, the measurement point is in the near field of the floor plate. The reason coupling does not occur in the far field is that adjacent zones of the plate are in antiphase, while the room mode shape is symmetrical, so sound radiated from one zone is cancelled by the antiphase radiation of the adjacent zone. In the near field, however, this cancelling involves volume flow along the path between the two zones and resulting finite sound pressure.

Ignoring the near-field effect, the sound pressure at a point x, y, x in the room, for a shape function for a simply supported plate, is then given by[11,12]

$$\frac{p_{x,y,z}}{v_{x,y}} = \frac{i\omega\rho c_o^2}{\pi^2 L_z}\left[\sum_{p,q,r=0}^{\infty} \frac{\Lambda_{p,q,r}(-1)^r \phi(x,y,z)\left[\sum_{m,n=1}^{\infty} \frac{mn\left[(-1)^{m+p}-1\right]\left[(-1)^{n+q}-1\right]}{(p^2-m^2)(q^2-n^2)\psi(x,y)}\right]}{\left[\left(\omega_{p,q,r} + i\delta_{p,q,r}\right)^2 - \omega^2\right]}\right] \tag{8.6}$$

where

$$\Lambda_{p,q,r} = \epsilon_p\, \epsilon_q\, \epsilon_r \tag{8.7}$$

and $\epsilon_p = 1$ for $p = 0$; $\epsilon_p = 2$ for $p > 0$; $\epsilon_q = 1$ for $q = 0$; $\epsilon_q = 2$ for $q > 0$; $\epsilon_r = 1$ for $r = 0$; $\epsilon_r = 2$ for $r > 0$;

The damping term δ is related to the reverberation time

$$\delta = \cong \frac{6.9}{T} \tag{8.8}$$

and the reverberation time is given by

$$T \cong \frac{0.161}{\frac{\epsilon_p \alpha_{y,z}}{L_x} + \frac{\epsilon_q \alpha_{x,z}}{L_y} + \frac{\epsilon_r \alpha_{x,y}}{L_z}} \tag{8.9}$$

where $\alpha_{y,z}, \alpha_{x,z}, \alpha_{x,y}$ are the average absorption coefficients of the room surfaces in the y, z x,z and x,y planes. It should be noted that for constant α the reverberation time is dependent on the mode number.

Where the measurement height $z < \frac{L_z}{m}$, $m + p$ in the numerator should be replaced by 1 and where $z < \frac{Ly_z}{n}$, $n + q$ in the numerator should be replaced by 1. When $m = p$ or $n = q$, the integral takes a different form, and $m + p$ or $n + q$ should be set to zero, avoiding the singularity.

These approximations are only valid for small values of α,which are likely to be the case at low frequencies in most rooms. In such cases, wall absorption is not due to a lining of soft material, but due to transmission of sound through the wall. This is a function of the admittance ratio of the wall.[13]

In the case of sound pressure in a room due to vibration from an underground railway, it is usual to measure both the floor vibration and the room sound pressure level at discrete points, not necessarily with the same x and y co-ordinates. It is clear from Equation (8.2) that the measured amplitude of the floor plate vibration will be dependent on the position of the measuring point, being at its maximum at an antinode. Likewise the sound pressure will be at its maximum in the corners of the room[14] and minimum at a node within the room. It is thus sensitive to both position on the floor and height above the floor. It is common practice to measure groundborne sound levels in rooms near the centre of the room, and this may result in a level being measured lower than the maximum which occurs in the corners. The position on the floor plate where the vibration amplitude is measured may or may not be at an antinode.

The results obtained with Equation (8.6) have been compared with the results obtained using a finite difference method, as implemented by the *FINDWAVE* program.

An important feature of the results is the presence of a series of antiresonances, and the prominence of the widely spaced room modes between 40 and 100 Hz.

In the case of a thick concrete slab, the lowest slab mode may be above the peak frequency in the incoming groundborne noise spectrum, and the slab may be in forced vibration at mode 0,0, that is, the entire building may be vibrating vertically as a piston. In this case, the wall is coupled with the room at mode 0,0,1 with a greater efficiency by a factor of $\frac{\pi^2}{4}$ or 7.8 dB. That will mean the walls are also in vertical vibration, in phase with the floor, and the slab that forms the ceiling of the room, if of similar construction, will be vibrating in phase with the floor slab. In so doing, it will be radiating downwards a similar pressure field in antiphase to that radiated by the floor, and the room shape function becomes

$$\phi(x,y,z) = \cos\frac{p\pi x}{L_x}\cos\frac{q\pi y}{L_y}\left\{\cos\frac{r\pi z}{L_z} - \cos\frac{r\pi(L_z - z)}{L_z}\right\} \tag{8.10}$$

This shape function would also occur in a tall building with similar floors.

It can be seen from Figures 8.3–8.6 that the relationship between sound pressure and velocity levels strongly depends on the relationship between the frequencies of plate modes in the floor, and those of room modes. If a plate mode should correspond with a room mode, then $L_v - L_p$ may be less than

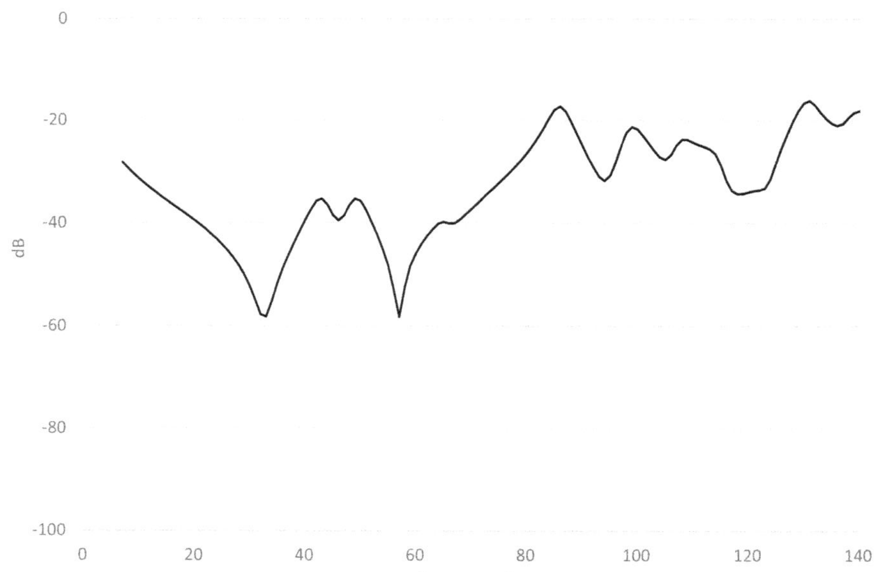

Figure 8.3 $L_p - L_v$ for a room 3.5 m × 4 m × 3 m high with reverberation time of 0.5 s, 1 Hz bandwidth.

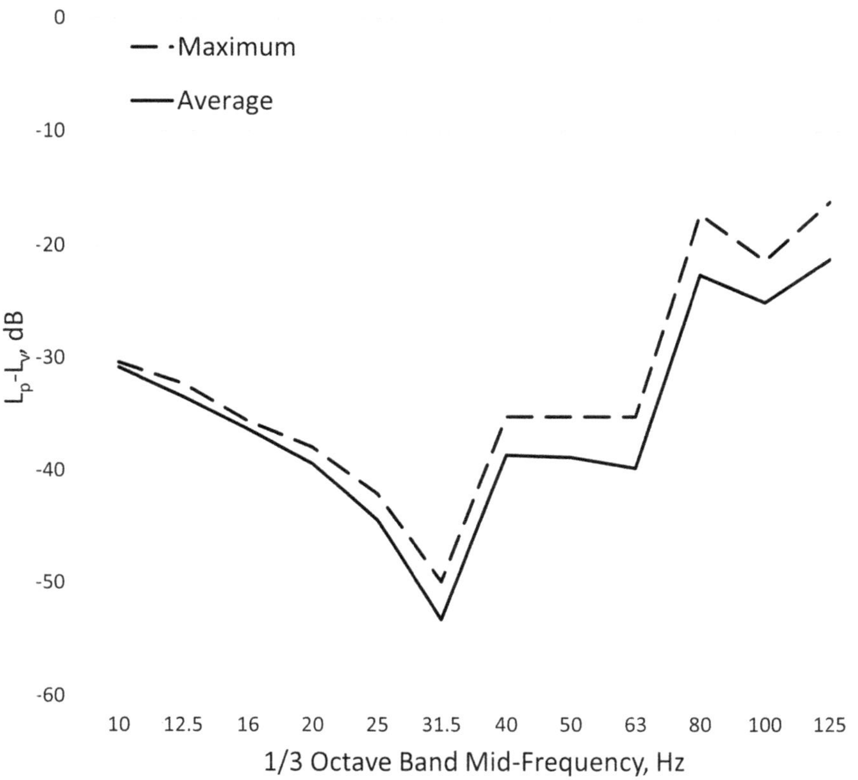

Figure 8.4 $L_p - L_v$ for a room 3.5 m × 4 m × 3 m high with reverberation time of 0.5 s, 1/3 octave bands.

20 dB, but, if it falls at the frequency of an antiresonance, it is very much greater. Depending on the approach taken to uncertainty, one may either take the worstcase, or alternately take a probability approach and use the mean square in third octave bands, as in Figure 8.4.

These results may be adjusted for a reverberation time T other than 0.5 s by adding the correction $20\log_{10}\frac{T}{0.5}$. This is a worst case applying strictly only to frequencies of room modes. If the floor plate can be assumed to be simply supported, then the results may be adjusted for other locations in the room using the correction

$$20\ \log_{10}\frac{\sin\dfrac{m\pi x}{L_x}\ \sin\dfrac{n\pi y}{L_y}}{\cos\dfrac{p\pi x}{L_x}\ \cos\dfrac{q\pi y}{L_y}\ \cos\dfrac{r\pi z}{L_z}} \tag{8.11}$$

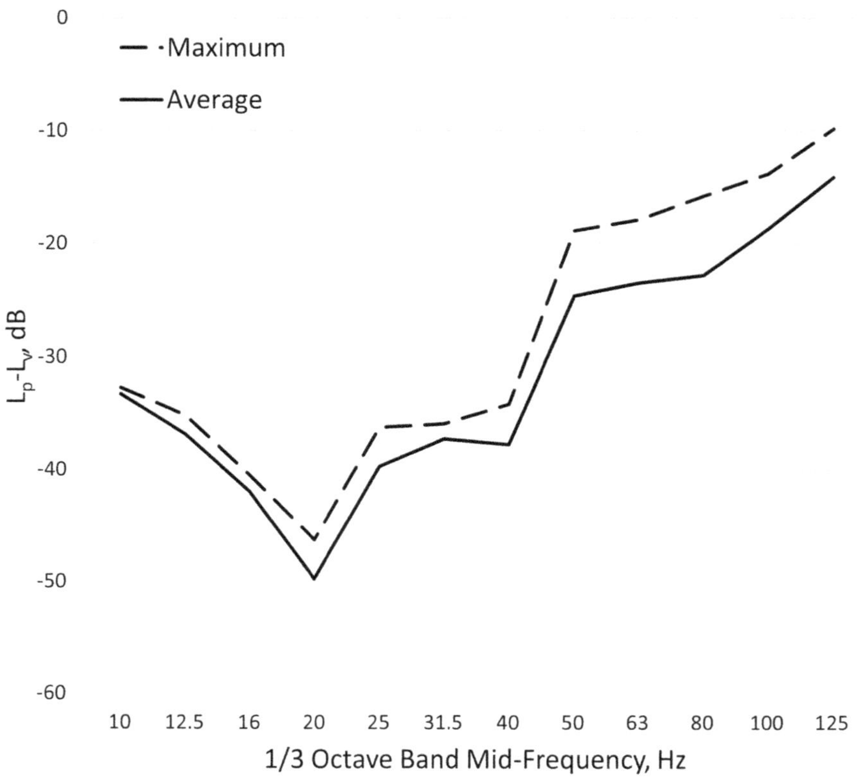

Figure 8.5 Maximum and mean square average of $L_p - L_v$ in third octave bands for room 6 m × 5 m × 3 m high with reverberation time of 0.5 s.

where x and y in the numerator are the vibration-measuring position and x, y and z in the denominator are the sound-measuring position.

The foregoing discussion applies to rectangular plates with simple support. Other edge support conditions do not have closed-form solutions to their shape functions, and the same is true of non-rectangular room shapes. Experimental studies of a large number of floors is reported by Auersch.[15] The effect of elastically supported edges and boundary absorption were considered by Asakura et al.[16] To examine the effect of irregular room shapes and non-simple edge supports, the *FINDWAVE* program was used to create a finite-difference-time-domain model of a complex building on piled foundations above an underground railway tunnel with resilient rail support. The ground floor slab is 250 mm concrete over a gas-venting cavity. The upper floors are constructed of 120 mm Cross-Laminated-Timber on a steel frame. Figure 8.7 shows plots of the floor vibration as A-weighted vertical velocity minus 27 dB and the corresponding room

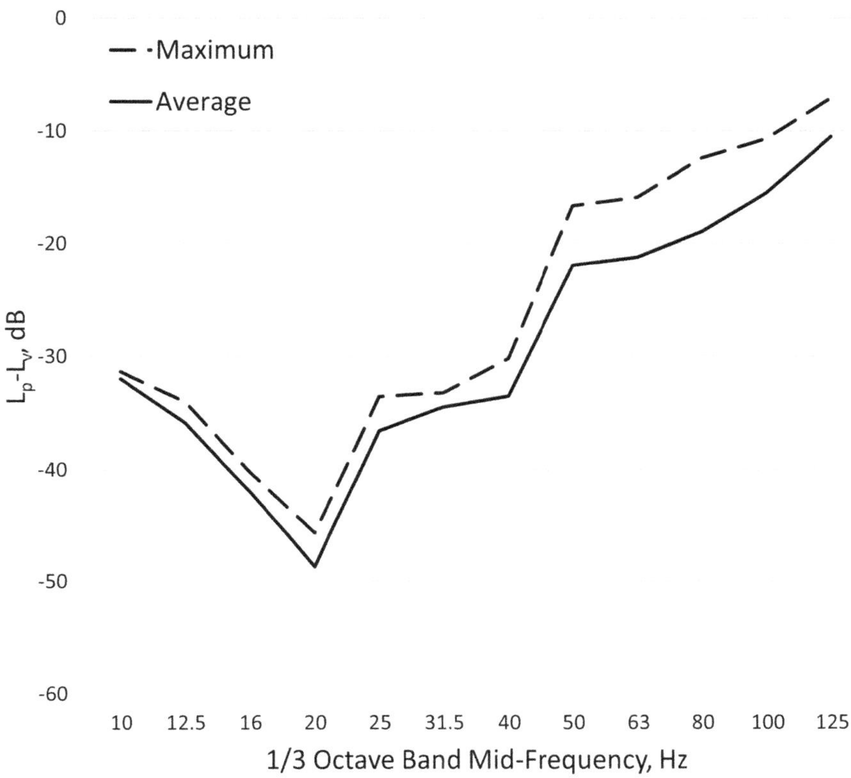

Figure 8.6 Maximum and mean square average of $L_p - L_v$ in third octave bands for room 6 m × 5 m × 2.5 m high with reverberation time of 0.5 s.

sound pressure level (SPL), assuming a reverberation time of 1 second. Figures 8.8 and 8.9 show comparisons between the spectra of L_v −27 dB (sometimes referred to as "pseudo-noise level") and SPL for the ground and first floors, together with the overall A-weighted values. Two comparisons are shown, one for locations of minimum levels in both cases and one for locations with maximum levels in both cases. For these cases the ground floor results show that the −27 dB correction is appropriate, but for the first floor there is wide divergence when the maxima are compared and L_v −27 dB at the location of the maximum L_v overestimates the SPL. In reality, measurements might be made at locations which may be near maxima for one metric and minima for the other, greatly increasing the uncertainty.

The relationship between floor velocity and room sound pressure level varies widely according to: (1) the relationship between measurement locations of the two quantities; (2) the room dimensions; (3) the reverberation

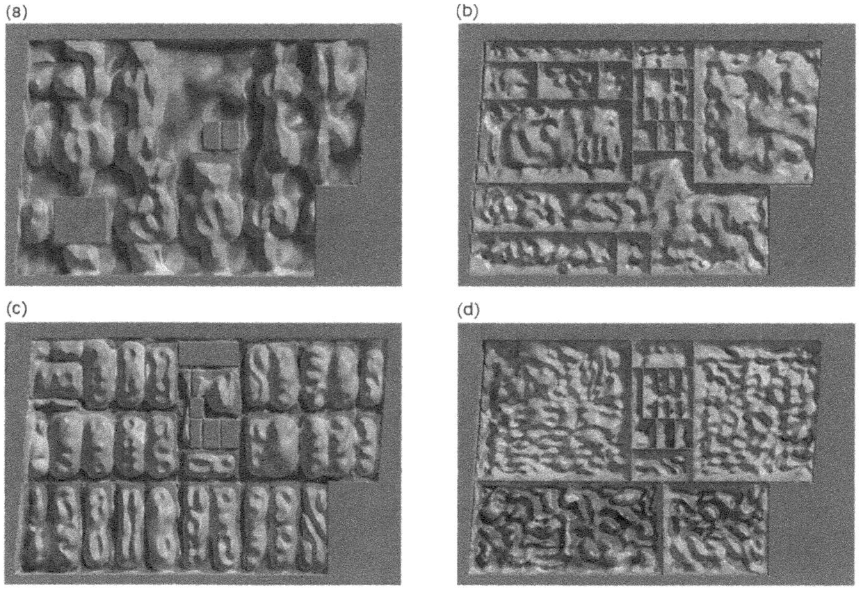

Figure 8.7 A-weighted vertical floor velocity −27 dB for ground floor (a) and first floor (c); and airborne sound level for ground floor (b) and first floor (d).

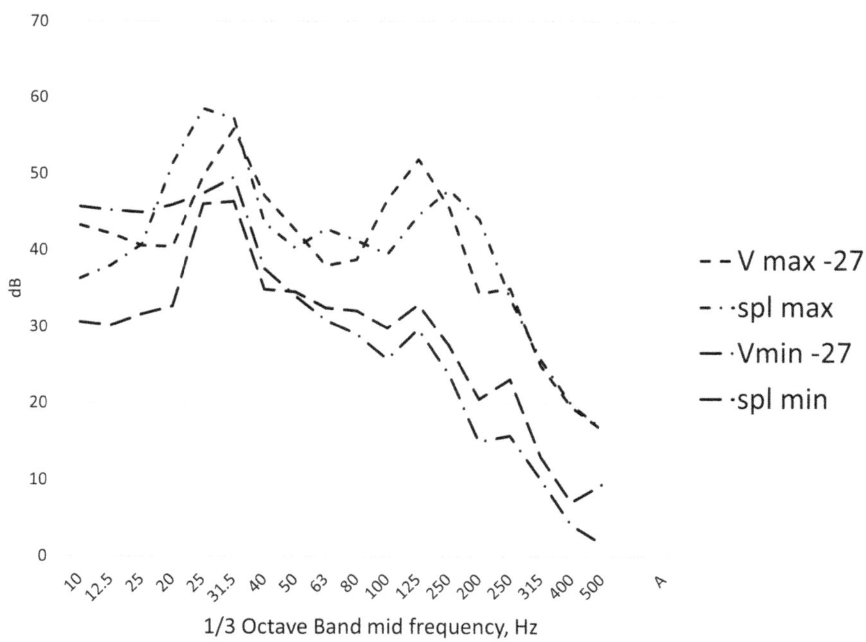

Figure 8.8 Comparison between L_v −27 and SPL for the ground floor.

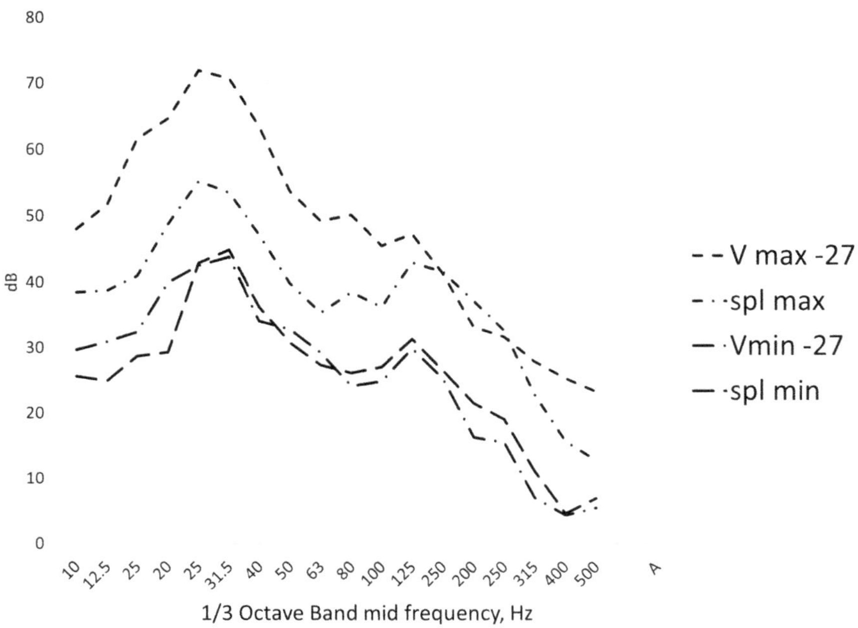

Figure 8.9 Comparison between L_v −27 and SPL for the first floor.

time of the room; and most importantly (4) the relationship between the frequencies at which plate modes in the floor occur and airborne room modes.

If, as is often the case with groundborne noise from underground railways, the floor vibration is concentrated at a single or small number of modes, slight differences between the match/mismatch between floor plate modes and airborne room modes may change the result by the order of 10 dB. In a case where it is essential not to underestimate the sound pressure level in a room likely to result from known or predicted floor vibration, a worst-case approach should be taken in which the modes are assumed to coincide. Otherwise a 1/3 octave band average may be used. Equation (8.6) can be implemented numerically in a spreadsheet.

NOTES

1. Kurzweil, L.G. "Ground-borne noise and vibration from underground rail systems." *Journal of Sound and Vibration*, 66(3), 363–370, 1979. 10.1016/0022-460x(79)90853-8
2. Zapfe, J.A., Saurenman, H., and Fidell, S. "Ground-borne noise and vibration in buildings caused by Rail Transit." Web-Only document 48. [Online] http://onlinepubs.trb.org/onlinepubs/tcrp/tcrp_webdoc_48.pdf

3. Kouroussis, G., Pauwels, N., Brux, P., Conti, C., and Verlinden, O. "A numerical analysis of the influence of tram characteristics and rail profile on railway traffic ground-borne noise and vibration in the Brussels Region." *Science of the Total Environment*, 482–483, 452–460, 2014. https://doi.org/10.1016/j.scitotenv.2013.05.083
4. Vogiatzis, K. "Protection of the cultural heritage from under-ground metro vibration and ground-borne noise in Athens Centre: The case of the Kerameikos Archaeological Museum and Gazi Cultural Centre." *International Journal of Acoustics and Vibration*, 17(2), 2012. https://doi.org/10.20855/ijav.2012.17.2301
5. Villot, M., Guigou, C., Jean, P., and Picard, N. "Definition of appropriate procedures to predict exposure in buildings and estimate annoyance." RIVAS Report, D1.6 RIVAS, 2012.
6. Federal Transit Administration TCRP Report. "Ground-borne noise and vibration in buildings caused by rail transit." December 2009.
7. Leissa, A. "Vibration of plates." NASA-SP-160 (1969), reprinted by Acoustical Society of America, 1993.
8. Fahy, F., and Gardonio, P. *Sound and structural vibration, radiation, transmission and response*, 2nd ed., Elsevier, Amsterdam, the Netherlands, 2007. https://doi.org/10.3397/1.2741307
9. Morse, P.M., and Bolt, R.H. "Sound waves in rooms." *Reviews of Modern Physics*, 16(2), April 1944. https://doi.org/10.1103/revmodphys.16.324
10. Cremer, L., Heckl, M., and Petersson, B.A.T., *Structure-borne sound*, Spring-Verlag, 2005. 10.1007/b137728
11. Kihlman, T., "Sound radiation into a rectangular room. Applications to airborne sound transmission in buildings." *Acustica*, 18, 12–20, 1967.
12. Hopkins, C., Filippoupolitis, M., and Ferreira, N. "Prediction of low-frequency radiation efficiencies using normal mode approach and finite element methods." *Proceedings of the 22nd International Congress on Sound and Vibration*, Florence, Italy, 12–16 July 2015.
13. Morse, P.M. *Vibration and sound*, Published by the American Institute of Physics on behalf of the Acoustical Society of America, 1981.
14. Simmons, C. "Measurement of sound pressure levels at low frequencies ion rooms. comparison of available methods and standards with respect to microphone positions." *Acta Acustica united with Acustica*, 85(1), 88–100, 1998.
15. Auersch, L. "Building response due to ground vibration—Simple prediction model based on experience with detailed models and measurements." *International Journal of Acoustics and Vibration*, 15(3), 2010. https://doi.org/10.20855/ijav.2010.15.3262
16. Asakura, T., Miyajima, T., and Sakamoto, S. "Prediction method for sound from passing vehicle transmitted through building façade." *Applied Acoustics*, 74, 758–769, 2013. https://doi.org/10.1016/j.apacoust.2012.11.011

Chapter 9

Measurement of groundborne noise

There are several reasons why measurement in the field of groundborne noise may be required. Starting at the receiver, it may be necessary to measure sound in a sensitive room, caused by a source of vibration, either to assess its acceptability, or to check compliance with a specification or validate a prediction model. Measurements at the receiver may also be made in order to discover the modal response of the structure. In many cases, the building containing the sensitive rooms may not have been built, and it is necessary to measure vibration in the ground where a proposed building will be constructed in order to assist in predicting future internal levels of sound. Measurements of vibration in the ground may also be made to seek to establish a distance function for the vibration. They may also be made for the purpose of characterising the source of vibration, that is, measuring the source strength. Finally, measurements may be made in order to evaluate the effect of installing a vibration mitigation measure at the source, or between the source and the receiver. In all but the first of these cases, it will not be the actual airborne sound level that will be measured, but the vibration amplitude, in terms of one or more of the parameters that describe vibration rather than sound.

MEASUREMENT AND ANALYSIS PRINCIPLES

An important matter in the measurement and analysis of vibration and groundborne noise is the way in which the data are captured and analysed. This is one of the many areas in which there are differences between practice in airborne acoustics and in groundborne/structure borne acoustics.

In airborne acoustics sound pressure levels are the most commonly measured quantities, and often the root-mean-square (RMS) level is the measured parameter. All RMS measurements have to have an associated time constant, and in airborne acoustics the use of exponential averaging has become well established. Exponential averaging is performed by computing the running RMS value using

DOI: 10.1201/9781003618775-9

$$A(t) = \sqrt{\frac{1}{\tau} \int_{\xi=0}^{t} e^{\frac{t-\xi}{\tau}} a^2(\xi) d\xi} \tag{9.1}$$

where a is the instantaneous signal value varying with time ξ; τ is the time constant, normally either 0.125 for "F" time weighting and "1.0" for "S" time weighting.

When vibration is being measured for the purpose of deriving levels of groundborne noise, exponential averaging is appropriate, with the time constant selected to accord with that used in the assessment criterion. In such circumstances, it may also be appropriate, having measured the signal as velocity or integrated an acceleration signal into velocity, to apply A-weighting, as used in airborne acoustics.

The A-weighting function is

$$W_A = 20 \log_{10} \left(\frac{A_1 A_2 f^4 F_4^2}{\left(f^2 + F_1^2\right)\left(f^2 + F_4^2\right)\sqrt{\left(f^2 + F_2^2\right)\left(f^2 + F_3^2\right)}} \right) \tag{9.2}$$

where f is frequency in Hertz, and

$A_1 = 1.007152$
$A_2 = 1.249936$
$F_1 = 20.598997$
$F_2 = 107.65265$
$F_3 = 737.86223$
$F_4 = 12194.217$

It is important to bear in mind that, while it is common practice to apply A-weighting from tables of 1/3 octave band values, at the low frequencies often involved in groundborne noise this can introduce significant error, if the signal is narrow band, because the value of the A-weighting curve at the signal frequency may differ by several dB from its value at the centre frequency of the 1/3 octave band in which it lies.

There are types of measurement in which A-weighting and exponential averaging are not appropriate. These include the measurement of transfer mobilities.

There are two principal methods of signal analysis where exponential RMS averaging is not appropriate. Usually these are cases where frequency spectra are required. The first is the Fourier transform, normally implemented using an FFT algorithm, and the second is digital filtering. In FFT analysis, an important consideration is windowing of the signal to remove spurious effects, due to the fact that the FFT algorithm treats a finite signal as if it is repeated indefinitely, so a step between the beginning and end of the signal introduces spurious components. The most common window applied is the Hann or Hanning window

$$W_h = \left(1 + \cos\left(2\pi\left(1 - \frac{N}{2}\right)/N\right)\right)/2 \tag{9.3}$$

where N is the number of samples in the window. For a signal which is an impulse that has not fully decayed before the end of the window is reached, a half-hanning window can be used, in which W_h is applied only to the part of the signal from $N/2$ upwards.

The important consequence of using a window is that the weighting causes energy to be lost from the signal, and compensation for this is required. For a stationary signal, the compensation comes from Parseval's theorem, and the frequency components should be multiplied by 8/3 to compensate for "leakage" caused by the Hanning window. For signals that vary significantly through the duration of the window, the correction required is less, but cannot be precisely determined.

When analysing an impulse using FFT, it is important to remember that the FFT algorithm assumes that the signal is repeated endlessly, and this introduces spurious effects.

MEASUREMENT OF AIRBORNE SOUND IN A ROOM

The first consideration is that groundborne noise caused by a vibration source tends to be of restricted bandwidth, as it is rare for significant levels of groundborne noise to occur above a frequency of about 1 kHz. This means that when selecting a microphone type (e.g., free-field, random incidence or pressure), there is negligible difference between their frequency responses in the relevant range.

However, whereas it is common to treat acoustic fields in rooms as diffuse fields, this assumption becomes less true the lower the frequency. In the frequency range where groundborne noise commonly occurs, particularly if it is due to a source such as an underground railway, the principles of modal acoustics have to be applied. This means that sound in the room is likely to consist of a system of nodes and antinodes associated with each of a sequence of resonant modes based on the geometry of the room. This subject was discussed in Chapter 8. The consequence is that the sound level in the room will be very dependent on measurement position. As Figure 8.2 of Chapter 8 shows, the commonly used measurement location being a position near the center of the room will be the quietest place to be for the first three mode combinations, and then at mode 2,2,0 it becomes the noisiest place to be. This problem is not easily overcome when the source of the noise is not constant or repetitive. This may be the case with groundborne noise from underground railways, and measurement using a large number of microphones may be impractical. Figure 8.2 of Chapter 8 shows that a potential solution would be to measure in the corners of the room, but, although some bedheads may be close to a room corner, it may not be a good representation of the sound that room occupants hear.

The German Standard DIN 45672-4[1] offers an indirect method in which the secondary airborne noise is estimated from the floor velocity levels, measured near the centre of the room.

The relationship between floor velocity and sound pressure is given in DIN 45672-4 by

$$L_{peq,m}(f_{Tn}) = A_n + B_n \cdot L_{veq,m}(f_{Tn}) \tag{9.4}$$

where A_n and B_n are listed in Table 9.1.

When groundborne noise is measured directly as airborne sound in a room, while there may be circumstances where sound radiated from vibrating surfaces in the room is completely predominant, there can also be circumstances where sound from airborne sources contributes. If the source of the vibration giving rise to the groundborne noise is a railway running on the surface, for example, there can, of course, be a contribution from the airborne path into the room that will be included in the measurements made.

The contribution of background sound from sources unrelated to the groundborne noise to be measured can be separated out if it is reasonably continuous or repeatable, by making measurements with and without the presence of the source of groundborne noise. However, when the source of airborne noise is the same as the source of the groundborne noise, the task of separating the airborne and groundborne contributions is not straightforward. If, as is often the case, the airborne sound is entering the room via the windows, temporarily increasing the airborne sound insulation of the room may be possible, for example by blanking off windows with sound barrier mat, but that may be impractical. Alternatively, sound-level measurements outside the window may be made, and a calculation of the resulting interior sound level, taking account of the sound insulation performance of the

Table 9.1 Coefficients A_n and B_n for the DIN 45672-4 third-octave correlation method by floor design

	Concrete floors		*Timber joist floors*	
Frequency (Hz)	A_n *(dB)*	$B_n(-)$	A_n *(dB)*	$B_n(-)$
25	32.36	0.418	32.57	0.446
31.5	27.98	0.501	29.33	0.528
40	28.84	0.506	25.23	0.615
50	25.33	0.557	20.01	0.660
63	22.62	0.595	22.58	0.543
80	23.66	0.597	26.80	0.463

window specification, is a course of action. Similarly measurements of the vibration amplitude of radiating surfaces in the room may be made, and a calculation carried out to predict the likely level of secondary sound in the room radiated from those surfaces. Both these approaches are subject to a considerable level of uncertainty because of the necessity to consider modal acoustics in the room, rather than diffuse field acoustics, if the sound is of low frequency.

MEASUREMENT OF VIBRATION OF SURFACES

For the measurement of the vibration amplitude of surfaces, the first decision to make is the type of transducer to be used. Vibration transducers are normally either geophones or accelerometers. Other less common transducers are laser vibrometers, and when the quantity to be measured is not acceleration, velocity or displacement, force transducers and strain gauges may be needed.

The first difference between geophones and accelerometers is that the former measure velocity, and the latter acceleration. All transducers have frequency limitations, including the effect of internal resonant frequencies. Geophones have resonant frequencies at the bottom of the frequency range, which have to be damped or compensated. Accelerometers have resonant frequencies at the top end of their range, which is usually a frequency much greater than the frequencies of interest in groundborne noise cases. Geophones tend to be larger and heavier than accelerometers, sometimes substantially so. Of particular importance in either case is the sensitivity of the transducer and its level of self noise, because the amplitudes of interest in groundborne noise measurement are often of extremely low amplitude, and many accelerometers are not sensitive enough, and/or do not give a large enough signal-to-noise ratio to enable valid measurements to be obtained.

While there are cases where measuring in one axis only is sufficient, such as the vertical axis, it is desirable to measure in all three axes when practicable. Some transducers are designed to be triaxial, otherwise three transducers are necessary, installed so that each one is appropriately orientated.

Accelerometers need either to be connected to charge amplifiers, or may be charge-coupled line drive devices or integrated electronics piezo-electric accelerometers (CCLD or IEPE) which require a power supply, whereas charge-amplified piezo-electric accelerometers do not. There are differences in their perfomance, one of which is that a CCLD device may require a short time to settle into its steady-state condition, outputting a spurious signal immediately after being powered, an effect that does not occur with charge-amplified piezo-electric accelerometers.

The most critical matter is the means by which the transducer is attached to the surface whose vibration it is to measure. This can significantly affect the resulting measurements. Studies[2] into the influence of transducer mounting methods have shown potentially large errors from this effect.

When the surface to be measured is flat and horizontal, a gravity attachment may be possible, if necessary aided by a very thin layer of beeswax or adhesive. Beeswax and/or adhesive may also be satisfactory for vertical flat surfaces. It is clearly important that the compression stiffness of the adhesive layer is such as to give a natural frequency well above the frequency range to be measured.

If the surface is not suitable for beeswax or adhesive, common practice is to use a circular metal disc (referred to as a DIN plate, according to the German standard DIN 45669-2[3]) with three pointed or rounded supports. Pointed supports can sometimes be pushed through a carpet when measuring floor vibration. A cube can be screwed on top of the disc so that three accelerometers can be mounted, in order to make triaxial measurements.

The difficulties arise when it is necessary to measure vibration in a ground surface, which may be soil, made ground, rough concrete or other non-smooth material. A variety of techniques have been used by practitioners ranging from driving a spike (either tubular in section or cruciform) into the ground and attaching the transducers to the top of the spike; casting a small patch of plaster-of-Paris, in a shallow depression on top of the soil, and fastening the transducers to the top surface thereof; burying the transducers; or fastening the transducers to a concrete tile or small slab.

Figure 9.1 from Reference 2 shows the comparative errors associated with these methods.

Guidance on transducer mounting is also given in Section 4.3 of PD ISO/TS 14837-31:2017[4] and in ISO 4866:2010.[5] The latter advises with respect to "Coupling to the ground" "Where transducers have to be mounted in the ground, in order to minimise coupling distortion, they should be buried to a depth of at least three times the main dimension of the transducer/mounting unit. Alternatively, they can be fixed to a rigid plate with a mass ratio $m/\rho r^3$ of not more than 2, where m is the mass of the transducer and plate; r is the equivalent radius of the plate; ρ is the bulk density, in kilograms per cubic metre, or the soil."

In all cases, care is required to secure the cables connecting the transducers to the recording unit, ensuring that they are neither strained nor loose and protected from moisture. Clearly the orientation of the transducers is key, and careful records including photographs are essential, with care taken to ensure correct identification of the three axes where separate cables are used.

Many measuring systems make it possible to monitor the signal being measured through the use of headphones. This is a good way to check for noise or distortion in the measurement chain and to ensure that a good quality signal is being measured. While some systems are auto-ranging or have a very large dynamic range, others may necessitate the correct choice of measurement range, which may be different for each axis, and it is important to ensure that the signal is well above the instrumentation self-noise floor and below the upper limit of the measurement range, to avoid distortion.

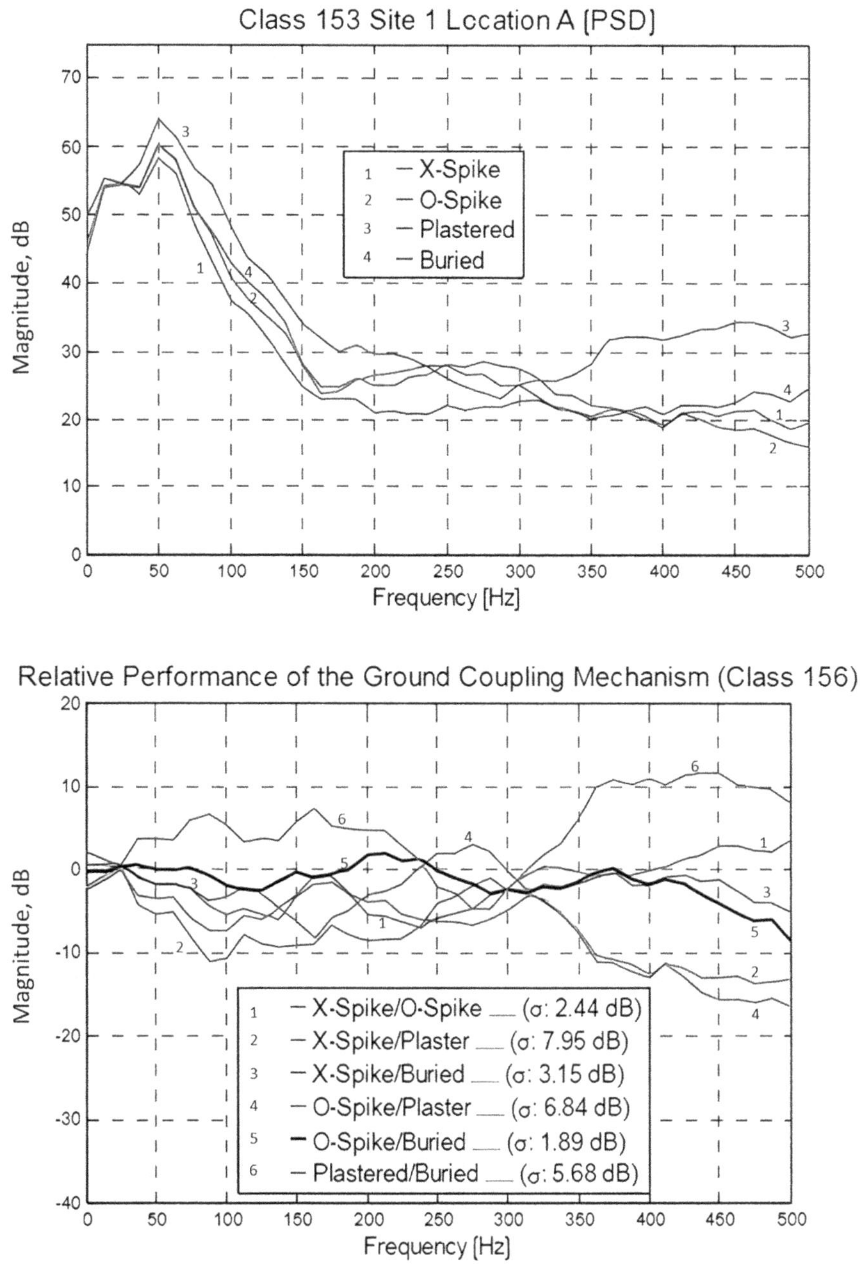

Figure 9.1 Spectra from ground-to-transducer coupling mechanisms, resulting from groundborne vibration induced by a passing train.

The range of a digital measurement system is fundamentally governed by the number of bits used in the analogue-digital conversion. Systems based on 16-bits can record a signal with an amplitude of $\pm$ 32768 (90 dB), although the poor resolution at the low end of the range will mean that effectively this gives an acceptable range of about 50 dB. The range is greatly increased with the use of 24-bit, 32-bit or 64-bit systems, although the data storage requirements are correspondingly greater. Data is usually encoded in WAV format and, although there is a standard format for WAV files, different products may employ significantly different implementations of the WAV format from the point of view of the encoding of measurement settings, sensitivities and other parameters.

Calibration of vibration measurement systems is an important topic. While vibration transducers are less sensitive to environmental and other influences that may cause drift from the calibrated sensitivity in sound transducers, that is, microphones, the use of a calibrator is desirable. For small, lightweight accelerometers, portable calibrators as convenient as microphone calibrators are available. For heavier transducers, including geophones, the calibrator may need to be a reasonably substantial device, effectively a shaker. As well as verifying the accuracy and correct functioning of a vibration measurement/recording system, a calibrator is valuable in providing an extra check on range settings and transducer sensitivities. Some systems require manual setting of ranges and sensitivities and, if these were not correctly noted at the time of set-up, results could be affected by a scaling error.

LINE SOURCE TRANSFER MOBILITY

An important measurement that may be carried out and used as part of the process of predicting levels of groundborne vibration and noise is the measurement of line source transfer mobility (LSTM).

The units of line source transfer mobility are $(\mathrm{m/s})/\left(\mathrm{N}/\sqrt{\mathrm{m}}\right)$, and thus it is a quantity that expresses the vibration velocity at a receiving point, which results from a linear input force. In energy terms it is $(\mathrm{m/s})^2/\left(\mathrm{N}^2/\mathrm{m}\right)$. The source is assumed to be infinitely long.

The test procedure[6,7] to measure transfer mobility basically consists of measuring the force (e.g., from an impulse such as a hammer blow or weight drop) applied to the ground and the response at a sequence of distances from the source. The test, therefore, involves creating vibration pulses that travel from the source to the receiver via the same path that will be taken by the vibration from the source being assessed. The transfer mobility is the relationship between the input force and the ground-surface vibration.

The measurement procedure is to generate the input force in a way that facilitates measurement of the force, for example, by applying a hammer or dropping a weight onto a force transducer or using a hydraulic or electromagnetic exciter. The responses of the force transducer and the receiver accelerometers or geophones are recorded for analysis.

When the procedure is applied to underground railways, the force must be located at the approximate depth of the subway. This can be done by drilling a borehole and locating the force transducer (that may need to be waterproof) at the bottom of the hole.

Signal-processing software is then used to calculate the transfer function and coherence between the force and vibration signals. Transfer functions should include the average of a number of impulses that can be averaged to provide signal enhancement to characterise the transfer function with sufficient accuracy. Signal enhancement is particularly important when the vibration transducer is more than 30 metres from the source.

Transfer mobility may also be measured using maximum-length sequence (MLS) force impulses. The transfer function can then be computed from the MLS forces and measured vibration. The MLS measurement method uses a pseudo-random binary sequence as the signal and has the advantage of increasing the signal-to-noise ratio of the measurement.

Two alternative methods can be used to estimate line-source transfer mobility. The first consists of using lines of transducers and the second consists of a line of impact positions. The second method is best for detailed assessment of specific vibration paths or specific buildings. The vibration transducers are located at specific points of interest, and a line of impacts is used. The point-source results are summed using Simpson's rule for numerical integration to directly calculate line-source transfer mobility. This is a considerably more direct approach than is possible with lines of vibration transducers.

Simpson's rule is a method of integration of a sequence of n discrete values between a and b:

$$\int_a^b f(x)dx \approx \frac{\Delta x}{3}(y_0 + 4y_1 + 2y_2 + 4y_3 + 2y_4 + \cdots + y_{n-1} + y_n) \qquad (9.5)$$

where $\Delta x = \frac{b-a}{n}$ and n must be even.

There is a method of computing the error associated with the discrete nature of the integration approximation, but, in the case of line source transfer mobilities, applying it does not make a significant difference to the result.

FORCE DENSITY

Force density, in units of $N/\sqrt{m}$ cannot be measured directly and has to be inferred from measurements of transfer mobility and train vibration at the same site. Force density can be inferred from line-source transfer mobility. The force density for each 1/3- octave band is then simply LF = Lv – Lstm where LF is the force density, Lv is measured train ground-borne vibration, and Lstm is the line-source transfer mobility.

It has to be borne in mind that the force created by an impulse is dependent on the impedance of the point receiving the force input. If the LSTM is obtained by applying impulses at the base of boreholes, the driving point impedance of the base of a borehole is not the same as, for example, the impedance of a tunnel wall. When using LSTMs for the purpose of predicting vibration from a future source such as a train, it is necessary to compute or numerically model the impedances for the two cases, the first being the measured location of the force input, the second being the support points of the future track, and to correct the results accordingly.

POINT SOURCE TRANSFER MOBILITY

If the source of interest is not linear such as a railway, but is more point-like, the determination of point source transfer mobility (PSTM) is a similar process to that for LSTM except that it is not, of course, necessary to integrate transfer mobilities along a line of discrete measurements. The same issues arise, particularly the need to take account of differences between the driving point impedance of the test point and that of the support of the future source of vibration.

A method of measuring driving point impedance is to measure simultaneously the force from the excitation using a load cell or force transducer and the acceleration of the mass used to generate the force. In the case of a hammer blow, instrumented hammers are available which incorporate both accelerometers and force transducers. The driving point impedance is obtained from complex frequency transforms of the two signals, and the impedance is the complex ratio of force to velocity (i.e., acceleration divided by angular frequency).

DISPERSION CURVES

If the process described above for measuring the point source transfer mobility is carried out using two or more phase-matched simultaneous transducers at a sequence of surface receiver locations, it is possible to measure dispersion in the surface wave. Chapter 3 explains that while Rayleigh waves on the surface of a homogeneous isotropic halfspace are non-dispersive, that is, they have the same phase speed at all frequencies, wherever the halfspace is layered, or has elastodynamic properties that vary with depth below ground, the effect is to create dispersion in which the Rayleigh wave speed is frequency dependent. Measurement of dispersion can be used to derive a profile of soil properties as they vary with depth.

Dispersion can be measured from the phase shift as a function of frequency as recorded for surface waves propagating between each receiver pair.[8] The phase velocity is calculated at each frequency using the relationship $c_r = (\omega d/\phi)$ in which c_r is the phase velocity in m/s, ω is the angular frequency, ϕ is the phase angle in radians (at frequency ω), and d is the distance between the receivers in metres. From this calculation, a plot of phase velocity versus frequency, called an individual dispersion curve, is generated for each receiver pair. This procedure is repeated for all source-receiver spacings used at the site and involves significant overlapping in the dispersion data between adjacent receiver pairs. Dispersion plots may alternatively be presented in terms of wave number (ω/c_r) as a function of frequency. The phase angle is determined from complex frequency transforms of the two receiver signals. Each transform can be expressed as magnitude and phase, and the phase angle can be extracted as a function of frequency. A dynamic soil profile can then be inferred by fitting computed and measured dispersion curves.

The method can be extended by the use of a string of many transducers, and by multichannel analysis of surface waves (MASW) more detailed soil profiles can be inferred from the resulting dispersion curves.

NOTES

1. DIN 45672. "Schwingungsmessung an Schienenverkehrswegen – Teil 4: Verfahren zur Ermittlung des sekundären Luftschalls."
2. D'Avillez, J., Frost, M., Cawser, S., l-Hamalawi, A., Fleming, P., Shields, P., and Skinner, C. "Practical Measurement of Railway Vibration." *Proceedings InterNoise 2010*, 2010.
3. DIN 45669-2:2025-02. "Measurement of vibration immission – Part 2 Measuring method." Deutsches Institut für Normung, Berlin. February 2025.
4. PD ISO/TS 14837-31:2017. "Mechanical vibration—Ground-borne noise and vibration arising from rail systems Part 31: Guideline on field measurements for the evaluation of human exposure in buildings." 2017.
5. ISO 4866:2010. "Mechanical vibration and shock—Vibration of fixed structures—Guidelines for the measurement of vibrations and evaluation of their effects on structures." 2010.
6. FTA. "Transit noise and vibration impact assessment." FTA-VA-90-1003-06, May 2006.
7. FTA. "Transit noise and vibration impact assessment manual." FTA Report No. 0123 Federal Transit Administration, September 2018.
8. Stokoe, K.H., Hwang, S., and Joh, S.H. "Spectral-analysis-of-surface-waves (SASW) testing to evaluate vs profiles at geotechnical and geological sites," *16th World Conference on Earthquake Engineering, 16WCEE 2017*, 2017.

Chapter 10

Prediction of groundborne noise

To predict groundborne noise it is necessary to characterise:

1. The source
2. The connection between the source and its immediate foundations or support system
3. The dynamic response of the immediate source support system
4. The interface between the source support system and the adjacent/surrounding structure or geology
5. The propagation function of the transmission medium or media between the source and the receiver
6. Coupling between the transmission medium and the receiver
7. The dynamic response of the receiver
8. Coupling between the structure/surfaces of the receiver and the air in the receiving space.

These eight stages are all discussed in previous chapters, from which it becomes clear that in each stage there is significant uncertainty, such that a manual calculation of received sound level due to a specific source can do no more than give a very approximate indication of the true result.

Algebraic solutions become extremely difficult and in many cases intractable. Usually such solutions are only possible when several major simplifications are made that introduce important differences between the model and the real-life case. Empirical models have their place, although large enough bodies of data, sufficient to give usable estimates of error, exist only for limited types of vibration, such as vibration from railways.

Equations are given in preceding chapters for many of the effects involved in the eight stages set out above, to enable an understanding of the underlying principles of the dynamic phenomena that occur, which is necessary for the interpretation of numerical results. In nearly all cases, however, numerical methods are the only realistic way of carrying out the required analyses.

DOI: 10.1201/9781003618775-10

MODELLING METHODS

Types of models

Numerical models have the attraction that, subject to the error associated with the discretisation that is always involved, they can yield exact solutions, if exact values for the input parameters are known. A numerical model is normally one of three types: finite-element models (FEMs), finite-difference-time-domain (FDTD) models, and combinations of either one with boundary-element models (BEMs). Finite-element models have the great advantage that there are many commercial packages available, and they are suited to complex physical systems and shapes. The FDTD models are not widely available and are normally restricted to Cartesian grids (although non-Cartesian grids are possible). Their advantage is that they are easy to code, and they model the vibration by simulating real life behaviour of elements of the model. Boundary-element models can be computationally efficient, particularly if combined with FEMs or FDTD models to solve the otherwise difficult integration over the surface of the source, but the required Green's function to represent the propagation characteristics of the medium between the source and the receiver is only trivial in cases of homogenous, isotropic material. Solutions for layered anisotropic ground become difficult. There are also algebraic calculation models which have been implemented in software, particularly the Pipe-in-Pipe model.[1,2]

Available software packages include the following.

Mefissto

Mefissto is software based on BEMs and FEMs developed at CSTB (Centre Scientifique et Technique du Bâtiment) in France. The user inputs any geometry as a set of contour segments (2D boundaries). It may contain any number of sub-domains where a sub-domain is characterised by the id numbers of one or two neighbouring media and the choice of a model (FEM or BEM) for each medium. Complex models with large buildings and multilayered grounds can be studied. Mefissto can be fully coupled with the VibraFer (CSTB) software that handles the source part for train/tracks systems.

Motiv

The MOTIV project (Modelling of Train-Induced Vibration) is a collaboration between the Universities of Southampton and Cambridge. The MOTIV software is a semi-analytical model developed in MATLAB for the prediction of ground vibration generated by surface and underground trains including the effects of propagation to the far field.

The MOTIV software can be used for assessing the performance of vibration mitigation measures at the track and/or the train vehicles.

The track/tunnel is modelled as an infinitely long structure, where the ground is modelled as a linear, layered and infinite soil medium. For surface railways, the software employs a model of a (ballasted, slab or floating slab) track that is longitudinally invariant and vertically coupled to the surface of the ground. Excitation is assumed to be due to the passage of individual wheel loads along the track (quasi-static loading) and due to dynamic interaction forces caused by irregularities of the wheels and tracks (dynamic loading). No coupling is made to the lateral shear stress at the soil surface. For underground railways, the software employs a model of an invariant (ballasted, slab or floating slab) track coupled to a circular tunnel embedded in the ground. The effect of a second tunnel can be included in the vibration prediction procedure. The software calculates the root-mean-square (RMS) vibration level of the vertical velocity at any selected point in the soil. The RMS vibration level is calculated for any given axle loads and wheel/rail irregularity profile.

For groundborne vibration the train is modelled as a linear, multiple, rigid-body system. In practice, due to the vehicle suspension, at frequencies above about 10 Hz usually only the unsprung mass of the wheelsets is important. Consequently, it is acceptable to neglect flexible modes of the car body. In MOTIV software, only vertical interaction forces are considered and only the vertical dynamics of the sprung masses of the train are included. Additionally, non-symmetric loading at the wheel/rail contact points occurs only at relatively short wavelengths (high frequencies; usually above 10 Hz). Thus, due to the presence of the suspension, the rolling modes of the car body and the bogie are not excited. Similarly, each wheel has only vertical displacement.

Each non-linear suspension component is assumed to be linear, which is valid for small motion amplitudes. Hysteretic, viscous or viscoelastic damping can be introduced and included in the stiffness elements of the vehicle suspension.

It is assumed that the wheel is always in contact with the rail. A linear Hertzian contact spring is included between each wheel and the rail, although, for the frequency range of groundborne vibration, inclusion of the contact spring does not influence the total response significantly. The software allows the calculation of ground vibration for any given train speed and rail roughness profile. The wheel irregularities are neglected, so that all irregularities are assumed to be on the rail surface. The rail roughness profile can be imported in the software or selected from an available library of theoretical and measured profiles.

In MOTIV software the ground is modelled as a linear, layered and infinite soil medium, that is, an elastic layered half-space. The parameters used for defining each layer of the ground are the thickness of the layer, the material density, the dilatational (p-wave) speed and shear (s-wave) speed, and the damping loss factor.

Pipe-in-Pipe

The Pipe-in-Pipe model was originally based on two concentric pipes to represent a cylindrical tunnel wall embedded in a full space. The inner pipe represents a tunnel wall and the outer pipe, with external radius set to infinity, represents the surrounding soil. The tunnel wall is, thus, a thin cylindrical shell and the surrounding soil is modelled using the elastic continuum theory. The tunnel and soil are uniform in the longitudinal direction and in the circumferential direction. A floating slab track may be coupled to the model by formulating the governing equations in the wavenumber-frequency domain.

An infinitely long train is represented by an infinite number of masses, representing its axles, and roughness is considered as the source of excitation. The power spectral density (PSD) of the displacement at a point in the soil due to dynamic forces at the wheel-rail interface is calculated with the dynamic forces assumed to be due to a white-noise rail-roughness.

Rigid bedrock below the tunnel is accounted for using the mirror-image method, such that vibration at the receiver is attributed to direct field from the source and reflection field from the boundary. The reflection field can be accounted for by removing the boundary and placing another source at the same distance from the removed boundary.

ElastoDynamics toolbox

The ElastoDynamics Toolbox is a MATLAB toolbox to model wave propagation in layered media based on the direct stiffness method and the thin layer method, presented by[3,4] Kausel and Röesset. It is used to solve a variety of problems associated with wave propagation in the soil, including site amplification, the computation of dispersive surface waves in layered soils, and the calculation of the forced response of the soil due to harmonic and transient loading.

Finite-element models (FEMs)

An advantage of the finite-element method is that there are open-source code packages, for example using Python, which make the method very accessible. However, to create an FEM for a specific groundborne noise analysis case requires substantial knowledge and experience of the creation of FEMs.

As the name suggests it involves first the discretisation of the model into elements. The relationships between these elements are then established using multi-degree-of-freedom system methodologies. While FEM accuracy improves with an increased number of elements, computational complexity and time requirements also increase in proportion. The major outputs sought from simple dynamic analysis of structures (modal analysis) are the modal frequencies, mode shapes and modal mass.

FINDWAVE®

FINDWAVE® is a finite-difference-time-domain (FDTD) model that computes in the time domain, in three dimensions, the vibration of structures and the surrounding air resulting from excitation by forces caused by, for example, the movement of rail vehicles. It also has an optional boundary-element module which assists in predicting sound-pressure levels at large distances from the radiating structure, without the need for an excessively large model. It is a very large programme using substantial computer memory with a runtime which may be several days for the largest and most complex models. It is optimised for an Intel 16-core processor, maximising parallel processing opportunities.

Regions to be modelled, which may be soil, rock and/or built structures surrounded by air, or underground tunnels with buildings above containing air, are represented in the model by a three-dimensional matrix of cells, each defined in terms of cell size, shear modulus, compression modulus, density and loss factor. Railways are represented by including the rails and support systems as cells, and by representing the rail vehicles wheel by wheel as lumped-parameter stacks of masses and springs, whose interface with the cell matrix moves according to the speed of the train. Excitation of the model is by the moving load effect of the wheels, and by the force resulting from the compression of the Hertzian contact spring between the rail head and the wheel tread as a result of wheel/rail roughness. A roughness profile is generated in the model, or in some cases read in from an external roughness profile file. There are methods of adjusting the results for alternative roughness assumptions. Output is in the form of the velocity or displacement in the time domain for the x, y or z direction for the selected cell(s), if the cells are solid, or of sound pressure, if the cells contain air. The time domain output files may be transformed into the frequency domain.

Finite-difference-time-domain (FDTD) models

The finite-difference approach[1] solves the wave equation by converting partial differential equations into discrete calculations. Despite the many types of wave propagation found in the transmission of vibration, there is only one wave equation for vibration in elastic media. It requires modification to include damping but otherwise is universal.

Stated mathematically, the wave equation for lossless propagation is

$$\mu\left(\Delta s + \frac{1}{1-2\nu}\operatorname{grad}\operatorname{div} s\right) = \rho\frac{\partial^2 s}{\partial t^2} \tag{10.1}$$

in which

$$s = \mathbf{i}\xi + \mathbf{j}\eta + \mathbf{k}\zeta \tag{10.2}$$

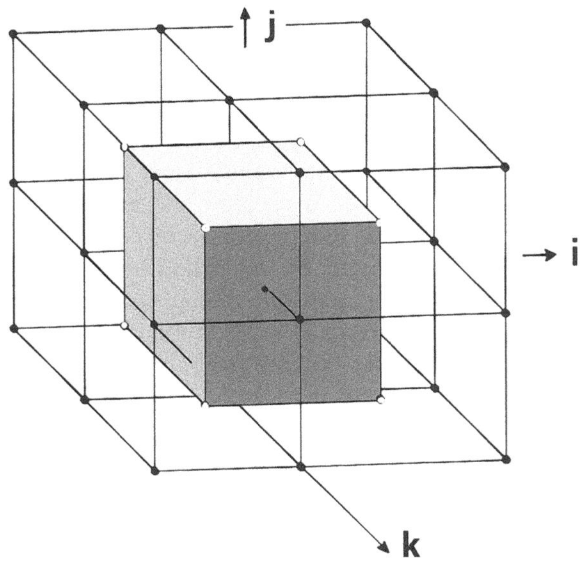

Figure 10.1 The basic finite-difference grid.

where ξ, η and ζ are displacements in three orthogonal axes; μ is the Lamé constant identical to the shear modulus; ρ is the density and ν is Poisson's ratio.

Figure 10.1 shows the finite difference representation of Equation (10.1). The region within the model is divided up into a three-dimensional grid of "shoe-box-shaped" cells, at each corner of which is a point shown as a black dot. Each cell, defined as one of the hexahedrons with four black dots as corners, is assigned a value for its shear modulus, compression modulus, density and loss factor, and for its displacement and velocity in each of the **i**, **j** and **k** directions. The vibration input to the model involves displacing, or adding velocity to, one of the "black dot" points.

At each time step, based on the displacement of the black dot points, the displacement of the white dot points that lie on a staggered grid halfway between the black dot points, is interpolated. These "white dot" displacements are then used to compute the strain on the grey cell in the centre. The strain is computed in terms of (1) the compression of the three pairs of opposite sides and (2) the shear of each face.

The force acting in each **i**, **j** and **k** direction is determined from the product of the strains and the relevant dynamic moduli. The centre of the grey cell is then accelerated in the resultant direction by the force divided by the mass of the cell, that is, its velocity increased by the product of the acceleration and the length of the time step. The displacement is then increased (or decreased) by the product of the velocity and the time step. This process is repeated for as many time steps as are required to achieve the necessary runtime, for example, one second.

The outstanding feature of this approach is that all wave types, whether shear waves, compression waves, bending waves, Rayleigh waves, Lamb waves or Stoneley waves will automatically occur simply as a consequence of the shape of the structure represented by the cells. A beam represented by a line of cells surrounded by air or a vacuum will exhibit bending waves, as will a plate represented by a two-dimensional layer of cells. A body of cells with a surface bounded by air or a vacuum will exhibit Rayleigh waves. In principle, shear waves occur independently of compression waves, but where there is a reflection caused by an obliquely incident wave on the boundary between two regions with different properties, wave conversion takes place. This occurs automatically in the finite difference grid.

The relationship between the general three-dimensional wave equation and equations for plates and shells when free surfaces are taken into account is explained in Reference 1.

Damping is not included in Equation (10.1), and additional manipulation is necessary to achieve damping in the model. There are basically three types of damping, viscous, friction and hysteretic. Viscous damping is caused by relative motion in viscous fluids, for example, water in a porous medium. Friction damping is caused by sliding, either of particles such as sand particles, or of surfaces such as the joints between steel beams. Hysteretic damping is caused by a lag between force and displacement and occurs in soils at large amplitudes.[4] The three types of damping have different frequency relationships. Viscous damping increases directly in proportion to frequency. Friction damping is independent of frequency.

Damping may be included in a finite-difference model by means of Boltzmann's expression[2] using relaxation functions.

$$\sigma(t) = D_1\epsilon(t) - \int_0^{\infty} \epsilon(t - \Delta t)\phi(\Delta t)d(\Delta t) \tag{10.3}$$

where $\phi(\Delta t) = \frac{D2}{\tau} e^{-\Delta t/\tau}$ is an after-effect function, D_2 is a constant and τ is a relaxation time. D_1 is a modulus, $\sigma(t)$ is stress and $\epsilon(t)$ is strain. By combining several after-effect functions with different values of D_2 and τ, any relationship between loss factor and frequency may be represented. In practice, three functions are sufficient to generate the frequency characteristics of most forms of damping, an example of which is given in Figure 10.2.

Absorbing boundaries

Some models require non-reflecting boundaries, and one of the early difficulties encountered in numerical modelling was the achievement of absorbing boundaries. That has been solved by the development of algorithms for the so-called "perfectly absorbing boundary".

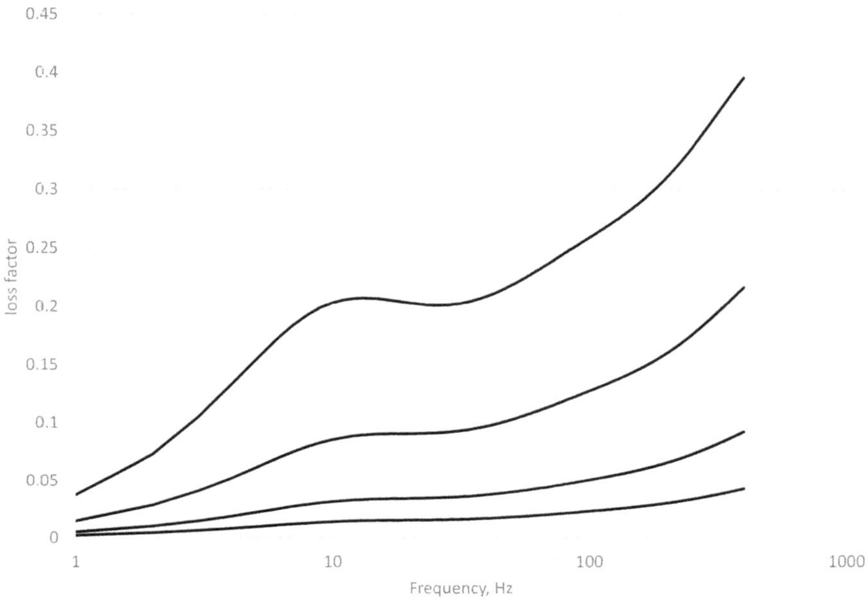

Figure 10.2 Example of frequency-dependent damping.

VERIFICATION AND VALIDATION

Verification of a numerical model involves comparisons between solutions for dynamic cases for which there is a correct closed-form solution, and modelled cases. Most numerical models can achieve agreement within a few percentage points.

Of greater importance is the relationship between modelled cases and real-life results. Provided that the model is verified as giving accurate results for cases with exact solutions, differences between modelled and measured results are attributable to differences between the modelled and actual parameters

The major input parameters in a numerical model are:

- The physical structure of the model
- Source characteristics
- Dynamic properties of propagating media.

The physical structure of the model

The physical structure of the model is as important as any other parameter, since in all numerical models discretisation is required, and this involves inevitable simplification. Furthermore, cell size must be several times shorter than the shortest wavelength encountered, and, since the need to minimise run times is important, there is always a desire to minimise the number of cells in a model. Because the different manifestations of wave motion that arise in a complex model all have different phase speeds, it is a non-trivial matter to determine the maximum cell size for the bandwidth studied.

Source characteristics

Clearly the output of a numerical model is a function of the input, and in some cases the input is not straightforward. For railways the exciting forces are well understood, and the input can be accurately modelled, and for impulse sources, such as percussive piling, the source is relatively simple, although in both cases the spectral content of the exciting force must be filtered to remove components outside the bandwidth of the model, itself determined by the relationship between cell size and wavelength. For footfalls on the floor of a steel framed building, force time-histories of a person walking are available. For rotating machinery, it may be necessary for a significant part of the model to be used to represent the source.

Dynamic properties of propagating media

The four parameters referred to above, shear modulus, compression modulus, density and damping affect the output of the model in the following ways. The moduli and density affect the wavenumber of eigenmodes in the system. This, in turn, will affect the degree of coupling between different parts of the system and influence the spectrum shape of the output signal. Furthermore, the effect of damping in attenuating travelling waves is a function of the wavelength of the wave, so, for a given loss factor, the attenuation in dB per metre is inversely proportional to the wavelength. The effect of layering in soil affects vibration propagation in several ways. Many soils exhibit a progressively increasing wave speed with depth, which has the effect of curving the propagation paths of body waves towards the surface.

Validation

Despite the potential for a numerical model to render a near exact solution to a case where all parameters are known, it is always the case that some or many parameters will not be accurately known. There are two ways of approaching the uncertainty resulting from unknown parameters. The first is purely an extension of the modelling exercise and involves first assigning probabilities to ranges of variation in the unknown parameters and second re-running the model many times with random variations in the unknowns. By this so-called "Monte Carlo" approach, a statistical treatment of the output can be made, with confidence limits determined.

The alternative approach is field validation. In many cases where modelling accuracy is critical, extensive fieldwork is carried out. This often involves geotechnical investigations, and measurements of transfer functions between boreholes or other excavations and the reception points are made, and modelled. An example of such as exercise is given in Figure 10.3.

An important consideration is the fact that the objective of most vibration prediction is comparison between a prediction and a requirement expressed in terms of a high-level index, such as $L_{A\,\max,S}$. As is evident from Figure 10.3 one of the effects of divergences between the model's representation of the world and the real world is frequently an error in the location of peaks in the spectrum. However, when the result is expressed in terms of the required assessment index, the influence of errors of that nature is small.

It is often necessary to predict the emission and propagation of waves in elastic media consisting of soil and rock. These media are not simply viscoelastic, because in almost every case they have some degree of porosity, very low in the case of rock such as granite, up to 30% or more in soils. In the latter case, the dynamic properties of the medium can be strongly dependent on the degree of water saturation, which can result in Poisson's ratio varying between about 0.3 and a figure approaching 0.5 for water. Among the uncertainties this can create is the fact that when only the shear modulus and compression modulus are known for the medium in its drained or partially drained condition, accurate calculation of the compression modulus applicable to field situations where saturation may be highly variable is not possible, with resulting uncertainty associated with attenuation rates,

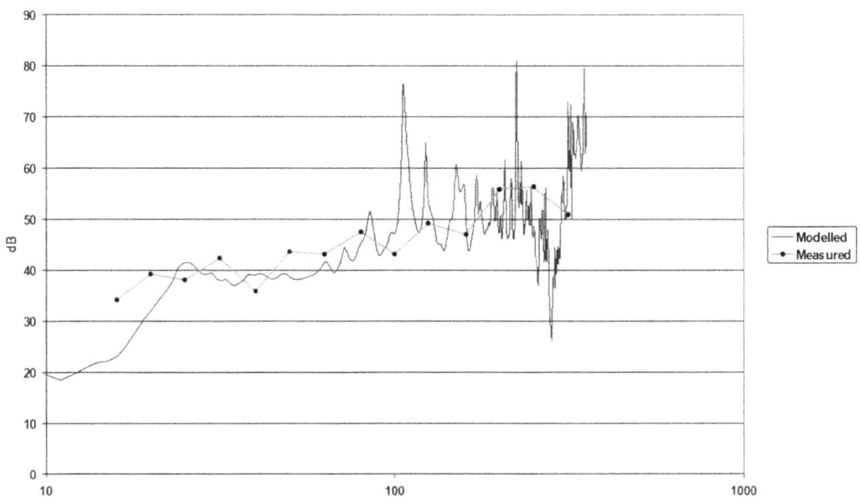

Figure 10.3 Comparison between modelled and measured transfer function between subsurface excitation points and ground level receptor.

and transmission losses at interfaces between different media. The behaviour of waves propagated in media with varying degrees of fluid saturation also explains some of the excess attenuation that is observed and is not accounted for by viscous, hysteretic and friction loss factors.

A method for predicting the emission and propagation of viscoelastic elastic waves is the finite-difference-time-domain approach, which is capable of adaptation so as to simulate the effect of fluid-filled porous media, by extension to include the two coupled wave equations originally developed by Biot for the dynamic behaviour of the elastic skeletal frame and the viscous fluid, whether water or air. Other authors have further considered the case of two immiscible fluids, such as oil and water as well as gas, but this case is not generally of interest in the case of groundborne noise and vibration.

The cases of groundborne noise and vibration to which the method can be applied involve significant vibration in the frequency range below approximately 500Hz, which make it possible to assume that the equations for the low frequency range may be used.

The finite-difference algorithm

Computation of wave propagation in elastic media using finite difference methods is an established procedure:

1. Compute pressure gradient for each grid cell
2. Compute shear force gradient for each grid cell
3. Accelerate $p(i,j,k)$ by $\Delta v = F/\rho\,\Delta t$ where F is the sum of the force derived from steps 1 and 2, ρ is the density assigned to the point and v is the point velocity.
4. Displace $p(i,j,k)$ by $\Delta x = \Delta v^{*}\Delta t$ where x is point displacement; Δt is one time step.
5. Repeat from step 1

The derivation process referred to in step 3 involves computing displacements at grid points $p(i + \frac{1}{2}, j + \frac{1}{2}, k + \frac{1}{2})$, and computing forces by interpolation at grid points $p(i,j,k)$. In the simple case of zero loss factor, the acceleration due to the force and pressure gradients is given (for the x-axis) by

$$\mu\left(\frac{\partial^2\xi}{\partial x^2} + \frac{\partial^2\xi}{\partial y^2} + \frac{\partial^2\xi}{\partial z^2}\right) + (\lambda + \mu)\left(\frac{\partial^2\xi}{\partial x^2} + \frac{\partial^2\eta}{\partial x\partial y} + \frac{\partial^2\zeta}{\partial x\partial z}\right) = \rho\frac{\partial^2\xi}{\partial t^2} \tag{10.4}$$

in which the differentiation is achieved in finite-difference form as

$$\frac{\partial\xi}{\partial x} \approx (x[i][j][k] - x[i-1][j][k])/\Delta x \tag{10.5}$$

Corresponding equations apply to the y- and z-axes.

NOTES

1. Hussein, M.F.M., Hunt, H.E.M., "A numerical model for calculating vibration from a railway tunnel embedded in a full-space." *Journal of Sound and Vibration*, 305(3), 401–431, 2007.
2. Hussein, M.F.M., Hunt, H.E.M., "The PiP model, a software application for calculating vibration from underground railways." *Proceedings of the Fourteenth International Congress on Sound and Vibration (ICSV14)*, Cairns, Australia, 9–12 July 2007.
3. Kausel, A., "Fundamental solutions in elastodynamics", Cambridge University Press, 2006
4. Kausel, E. and Röesset, J. M., Stiffness matrices for layered soils, *Bulletin of the Seismological Society of America*, Vol. 71, pp. 1743–1761, 1981

Chapter 11

Mitigation of excessive groundborne noise

Mitigation of excessive groundborne noise may be achieved in a number of ways, and the most fundamental method is to provide vibration isolation at source or in the path from source to receiver.

Vibration isolation normally involves the insertion of a resilient element between the source of vibration and its transmission path, or between the transmission path and a vibration-sensitive receiver. For example, resilient bearings may be inserted beneath concrete blocks supporting railway track, or a building may be constructed on base-isolation bearings. "Isolation" is not an accurate term as the inclusion of resilient elements does not prevent all vibration from passing through them. The extent to which the transmission of vibration is reduced by this method is a matter of impedances. The elementary case found in all textbooks covering the topic is that of the "mass on a spring", and for many decades the practice of installing machines on anti-vibration mountings has been commonplace. A first-order approximation to the reduction in vibration transmissibility achieved in this way is to regard the machine as a lumped mass, the anti-vibration mounting as a single spring and the foundation on which the mounting stands as infinitely stiff. It is then quite straightforward to calculate the force transmissibility of the bearing.

The impedance of the lumped mass is $i\omega m$, and the impedance of the anti-vibration mounting is $k/i\omega$. The impedance of the foundation is assumed to be infinite. The simplest approach is to regard the system as an electrical circuit with the mounting as a capacitor, shunted across the circuit between an inductance representing the lumped mass and a resistor representing the foundation.

Then, the ratio of the voltage across the resistor (i.e., the force on the infinitely stiff foundation, often referred to as the blocked force) to the voltage across the whole circuit (i.e., the force exerted on the lumped mass) is $1/(1 - \omega^2 m/k)$. The first conclusion is that when $k/m = \omega^2$, the result is infinity. The system is in resonance, and the energy of the source is stored in the mass and the spring as kinetic and potential energy, building up

DOI: 10.1201/9781003618775-11

continually as it has nowhere to go. When $\omega^2 m/k$ is equal to 2, the force on the foundation is the same as the force on the mass, above which point it becomes less. When ω^2 is much greater than k/m the force on the foundation is inversely proportional to ω^2, and, in terms of the decibel scale, that means a force reduction of 12 dB per doubling of frequency. That offers the prospect of large transmission losses at audible frequencies. For example the force transmission loss at an exciting frequency of 100 Hz in a system where the natural frequency is 10 Hz is 40 dB. In practice, there is normally damping present, which may be represented as a resistor in series with the capacitor representing the spring. The same equation applies in the opposite direction when the exciting force is the foundation, and the matter of interest is transmissibility into the isolated mass. This latter case is, of course, the simple representation of the base-isolation of a building. However, a building is not a lumped mass, and its impedance is not simply $i\omega m$. A building is a complex network of mass-like and spring-like elements. In the case of a reinforced building with concrete slabs (i.e., plates) supported on concrete columns (Figure 11.1).

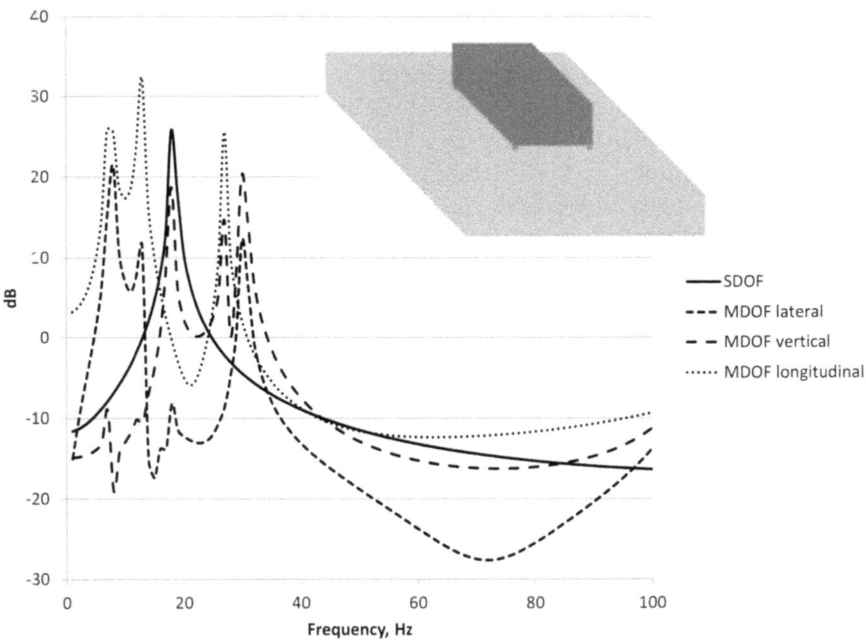

Figure 11.1 Velocity transmissibility of single-degree-of-freedom (SDOF) and multi-degree-of-freedom (MDOF) systems.

ISOLATION AT THE RECEIVER

Predicting the performance of base-isolation systems for buildings for the control of groundborne noise requires the use of a model of the dynamic behaviour of the building. Frequently the model used is that of a single-degree-of-freedom (SDOF) system consisting of a mass on a spring, and the classical equation for the transmissibility of the spring foundation is used as a means of predicting the behaviour of the base isolation system. Such a model necessitates an assumption that the building is a lumped mass m, and the input excitation for all the bearings of stiffness k is equal in magnitude and phase so that only the vertical motion need be considered. The classical equation leads to the conclusion that above a frequency of $\sqrt{2}f_n$ where f_n is the bearing natural frequency $\sqrt{(k/m)}/2\pi$, there is a reduction in the vibration amplitude of the isolated mass, which decreases at the rate of 6 dB per octave for a viscously damped system with critical damping and up to 12 dB per octave for an undamped system, with no upper frequency limit.

In practice, no real building satisfies these assumptions. It is common practice to derive the values for the mass supported by each bearing from the structural engineer's load calculations and divide the load in Newtons by the gravitational acceleration g (which engineers frequently simplify by assigning it the value of 10 m/s^2, so that load in Newtons is taken as 10 times mass in kilogrammes). This does not necessarily give the correct value for the mass above the bearing, as the load force can be due in part to strain in structural members, and not simply to the weight of the supported mass. Even a building which approximates a lumped mass may rock, pitch or roll at one of the coupled natural frequencies associated with those degrees of freedom. Few, if any, buildings approximate a lumped mass, and almost all have internal natural frequencies of columns, floorplates and other structures. Tall buildings, which are increasingly the subject of base-isolation designs, may behave as a stack of mass-spring systems with the floors as masses and the columns as springs.

THE DYNAMIC BEHAVIOUR OF BUILDINGS

Theory

The equation commonly used for calculating the performance of a base-isolation system using a SDOF model is[1]

$$T = \sqrt{\frac{1 + \left[2\left(\frac{f}{f_n}\right)\left(\frac{b}{b_c}\right)\right]^2}{\left[1 - \left(\frac{f}{f_n}\right)^2\right]^2 + \left[2\left(\frac{f}{f_n}\right)\left(\frac{b}{b_c}\right)\right]^2}} \tag{11.1}$$

where f is the forcing frequency, $f_n = \sqrt{(k/m)}/2\pi$ is the natural frequency of the mass-spring system and b/b_c is the fraction of critical damping. This equation is identical (11.1) to the equation for the force transmissibility of a vibrating mass mounted on a spring, except that for a case of base-isolation the transmissibility, T, is of displacement, velocity or acceleration and not force. Since groundborne noise is broadly a function of the vibration velocity of radiating surfaces, Equation (11.1) will be regarded as a velocity transmissibility equation in this chapter.

It follows that even in a case where a building behaves like a lumped mass and is excited only vertically by an input in-phase at all mountings, the velocity transmissibility is not necessarily the insertion loss of the bearings. The velocity of the bearing foundation will not be the same as it is with rigid blocks in place of bearings. For a given input force, the effect of inserting a low impedance element in the transmission path may be to increase the velocity, reducing the isolation calculated using Equation (11.1). However, this assumes that the source has constant force characteristics. The driving point impedance of the base of a mass-on-a-spring system, at resonance, is infinitely high for zero damping. Consequently, the velocity of the base is zero at resonance, because it cannot move against an infinite impedance. The steady-state transmissibility is infinite, however, because there will be some excitation of the mass at the start of the excitation, and its velocity divided by that of its foundation involves division by zero to give infinite transmissibility.

In almost all real cases, in addition to vertical movement, a supported mass will also be excited asymmetrically causing it to rock or pitch. These modes will have their own natural frequencies, each of them coupled to the vertical SDOF natural frequency. The effect of the coupling is to cause two new coupled frequencies in each case; one lower than the SDOF frequency and one higher. Lateral components in the foundation vibration will excite either or both the two lateral natural frequencies of the mass and may also excite the yawing mode (rotation about a vertical axis). Thus, there are potentially six natural frequencies associated with a lumped mass on bearings.

Most buildings are constructed as assemblies of plates and beams, including beams forming columns to support floor plates. The bearings are likely to be placed beneath columns. The transmissibility of the bearing system will be dependent on the driving point impedance of the structure "seen" by the top of the bearing. An infinitely long column standing on end on a bearing has a driving point impedance of $\rho c_L S$ where ρ is the mass density of the column, c_L is the longitudinal compression wave speed and S is the cross-sectional area of the column.

Vibration is not only transmitted through columns by means of longitudinal compression waves, but also by means of bending waves, if the motion of the foundation is not in-phase across the base of the foundation. Plates supported by columns will be excited in one or more of their plate modes as a result of both vertical and bending wave inputs from columns. Examples of the driving point impedance (and its inverse, mobility) of plates were shown in Figures 4.3, 4.4 and 5.5. These showed excursions of the order of as much as 60 dB, depending on damping, between resonances and anti-resonance at frequencies quite closely spaced.

The consequence of the existence of the significant number of mechanisms that cause a building to depart from the lumped mass assumption is that whereas the impedance of a mass is simply $i\omega m$ (where ω is the angular frequency, m is the mass in kilograms and i is $\sqrt{-1}$), the impedance of a building contains many resonances at which the transmissibility of the bearings will be significantly above the SDOF value, and anti-resonances at which it is less.

The precise behaviour of a base-isolated building is dependent on the source impedance. This is a function not only of the nature of the foundations below the bearings, but also on the coupling between the building and the soil or rock supporting it. A finite plate, resting on the top of an elastic halfspace, behaves like a mass on a spring,[2,3] with the mass being not only the building foundation structure, but also a volume of soil near it, and the spring constant being a function of the dimensions and the soil dynamic properties. In these circumstances, the SDOF mass-spring system becomes a mass-spring-mass-spring system, possessing two coupled natural frequencies, one below the foundation natural frequency, and the other above the supported mass-spring combination.

In contrast to the simple case of a rigid body resting on a half-space is a tall building with many floors supported on columns. Such a case has interesting dynamic properties.[4] For example, if the load on the ground floor columns is summed and considered as a single degree-of-freedom system with the ground floor columns treated as springs based on the compressive stiffness of the columns, it can be the case that the natural frequency of system is in single figures. Should it be necessary to consider base-isolation of such a building, the effect of inserting bearings below the columns would seem intuitively to be somewhat nugatory. However, the actual situation is much more complex than that of a single-degree-of-freedom system, as illustrated in Figure 11.2. The figure is a plot of the velocity transmissibility between the foundation and the first floor of a nine-storey building, showing the peaks caused by the SDOF resonances of the all the columns and masses of the slabs above. The upper curve is for an unisolated building, and the lower curve shows the effect of inserting resilient bearings below the lower floor level, having a natural frequency of 7 Hz calculated using the total mass of all the floors as if they were a lumped mass. Although the improvement in transmissibility is much less than the SDOF equation predicts, there is, nonetheless, an improvement of 10–15 dB. It is notable, however, that there is already a significant reduction in transmissibility in the unisolated case at frequencies above the natural frequency of the top floor on its columns, in this case, above 80 Hz.

Numerical modelling

The stack of springs and masses considered in Figure 11.2 is a major oversimplification, particularly in an asymmetric building structure. In order to study the effect of introducing resilient bearings in a tall building, a numerical model has been created using a finite-difference-time-domain (FDTD) software package.[5] The building in question is situated over an underground railway and has a number of deep basements and podium structures, above which is an 18-storey residential tower with irregularly shaped floors. In order to control groundborne noise from the underground railway in the residential apartments, the building is designed with resilient bearings at the heads of the columns and in the cores just below first floor level, being the first of the eighteen floors of the tower. The bearings were assigned stiffnesses based on a SDOF calculation, taking the entire mass of the building above the bearings such that they would have a natural frequency of 10 Hz.

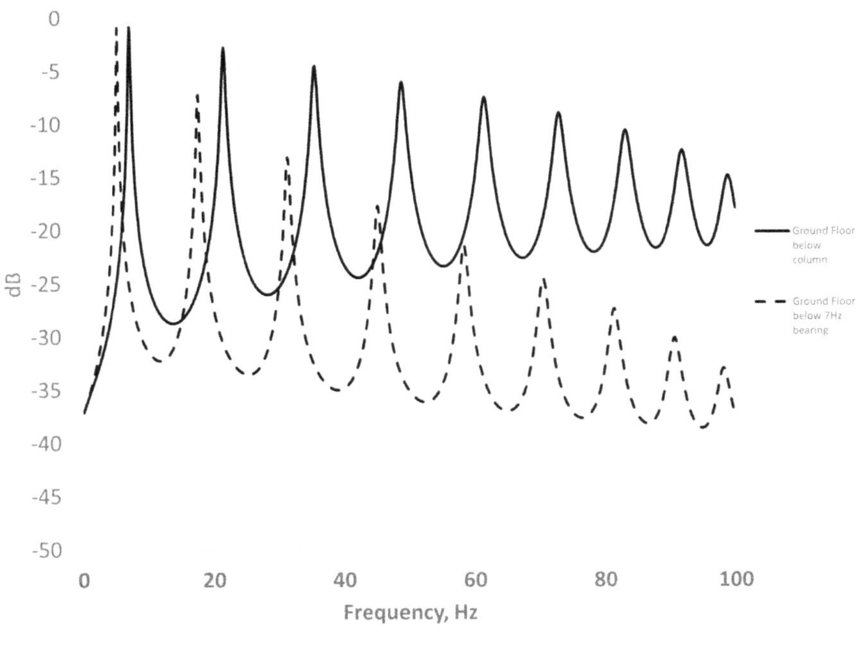

Figure 11.2 Coupled natural frequencies in a nine-storey building with and without 7 Hz base-isolation bearings inserted.

Many of the dynamic properties of a building are very difficult if not impossible to measure in-situ, and the advantage of a numerical model is that information can be output that would be unavailable for the actual building.

The driving point impedance was output from the model at the top of a bearing underneath a column line, defined as $\mathcal{F}\{F(f)\}*/\mathcal{F}\{v(f)\}*$ where $\mathcal{F}$ signifies the Fourier transform of the force, $F(f)$ and velocity, $v(f)$ and * signifies the complex conjugate. It is plotted in dB in Figure 11.3 along with the two idealised lumped-parameter impedances of $i\omega m$ and $\rho c_L S$. It can be seen that, while peaks in the spectrum reach the upper $i\omega m$ curve, at most frequencies the curve is closer to the lower $\rho c_L S$.

Figure 11.4 shows, in 1/3 octave bands, the velocity transmissibility relative to the base velocity of the building with the bearings in place, and Figure 11.5 is the velocity transmissibility relative to the base velocity of the building with rigid support. So it takes account of the effect of the bearings on the source below as well as the structure above. Of particular note is the absence of a peak at the SDOF bearing natural frequency, and the presence of two coupled peaks, one above and the other below the SDOF natural frequency, resulting from the coupling between the structure above the bearings and the foundations below. Comparison of Figure 11.5 with Figure 11.4 shows that when the change in the foundation velocity is taken

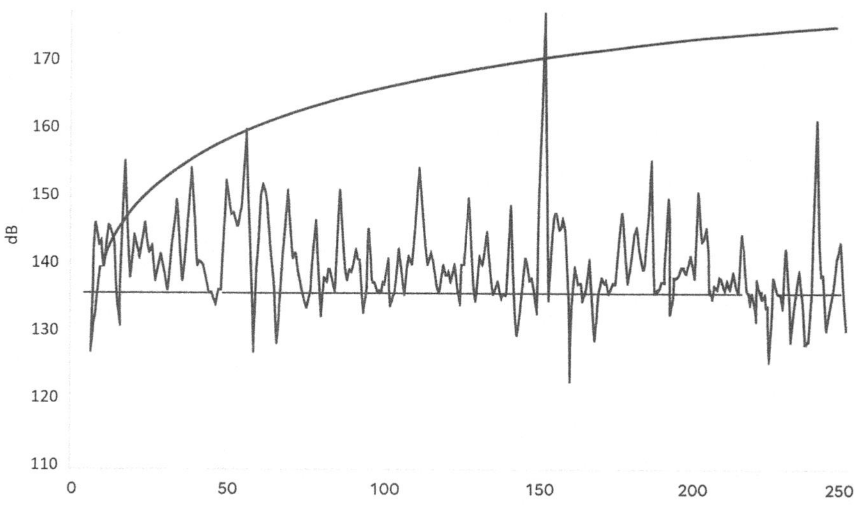

Figure 11.3 Driving point impedance of (1) column above bearing (middle curve) together with (2) an equivalent lumped mass (upper curve) and (3) an infinite column.

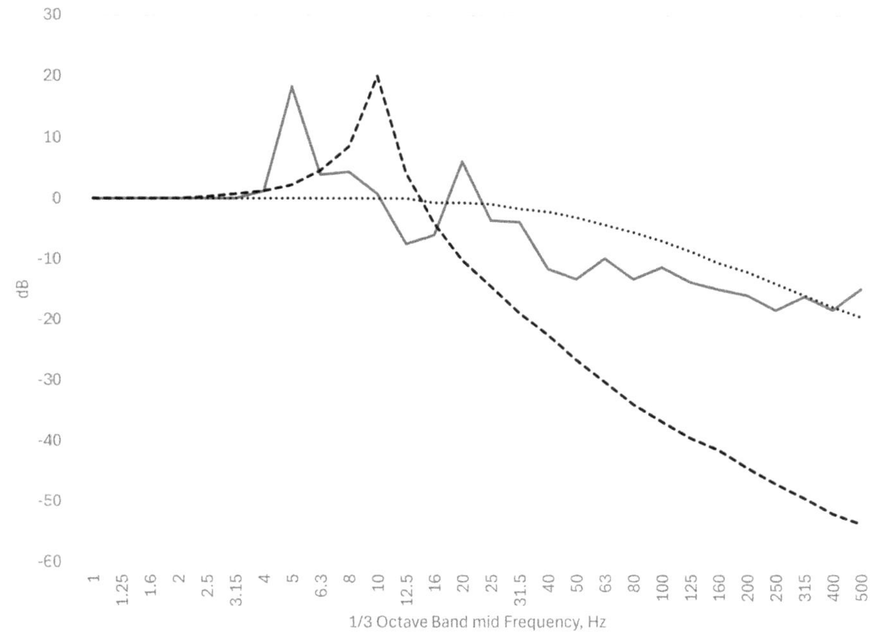

Figure 11.4 Velocity transmissibility of the 10 Hz bearings (1) from the FDTD model (middle curve) together with (2) an equivalent lumped mass (lower curve) and (3) an infinite column (upper curve).

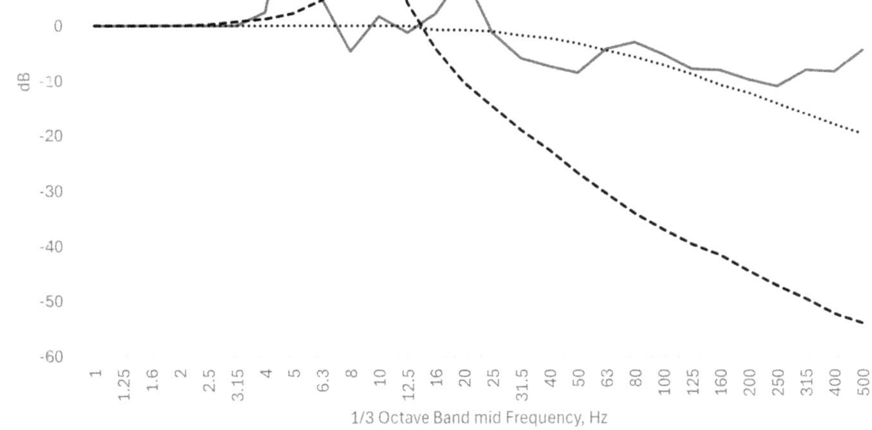

Figure 11.5 Velocity transmissibility of the 10 Hz bearings relative to the base in the unisolated building (1) from the FDTD model (middle curve) together with (2) an equivalent lumped mass (lower curve) and (3) an infinite column (upper curve).

into account, the actual transmissibility more closely resembles that predicted using the driving point impedance of an infinite column than a lumped mass.

STRUCTURAL DAMPING

Given that the amplitude of vibration in buildings, and thereby the level of sound radiated into spaces in the building, is strongly influenced by the modal response of building elements such as floor plates, the question arises whether modal responses can be reduced by added damping. Merely adding a viscoelastic layer to a floor will not be an efficient way to add damping, but if the viscoelastic layer is constrained by a plate, reasonable increases in damping can be achieved. This is because bending of the resulting composite causes not only bending and extensional strains but also shear of the viscoelastic layer. If the stiffness of the viscoelastic layer is small relative to that if the floor plate and the constraining layer, the loss factor of the composite panel is given by

$$\eta = \frac{\beta_2 YX}{1 + (2 + Y)X + (1 + Y)\left(1 + \beta_2^2\right)X^2} \qquad (11.2)$$

where X is the shear parameter

$$X = \frac{G_2}{k^2 H_2}\left(\frac{1}{E_1 H_2} + \frac{1}{E_3 H_3}\right)$$

and

$$\frac{1}{Y} = \frac{E_1 H_1^3 + E_3 H_3^3}{12 H_{31}^2}\left(\frac{1}{E_1 H_1} + \frac{1}{E_3 H_3}\right)$$

and E_1, E_3 = moduli of elasticity of the two elastic layers (N/m^2); H_1, H_3 = thicknesses of the two elastic layers, (m); $H_{31} = H_2 + (H_1 + H_3)/2$; G_2 = real part of the complex shear modulus of the viscoelastic layer; (N/m^2); $k = 2\varpi/\lambda$ = wave number (m^{-1}); $\lambda = \sqrt[4]{\frac{\overline{B}}{\rho_s}}\frac{2\pi}{\sqrt{\omega}}$; $\overline{B} = (B_1 + B_2)\left(1 + \frac{\overline{X}Y}{1 + \overline{X}}\right)$; $\overline{X} = X(1 + j\beta_2)$; β_2 = loss factor of the viscoelastic layer (dimensionless).

Equation (11.2) is plotted in Figure 11.6.

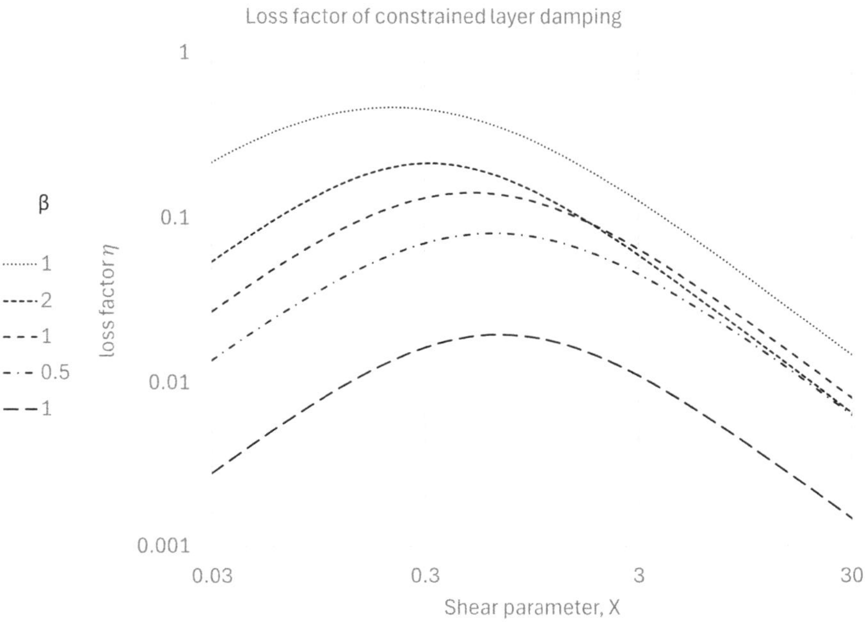

Figure 11.6 Loss factor of constrained layer damping.

ISOLATION AT THE SOURCE

Isolation of vibration (and therefore groundborne noise) at source is most commonly a solution when the source is a machine, and the machine may be a railway. There is a significant difference between the approach that can be taken in the case of a static source, such as a machine installation, and a moving source such as a railway.

Static sources

Vibration isolation of static sources has a long history, and information about the topic is found in most vibration handbooks. There are many manufacturers of anti-vibration mountings or vibration isolators, and it is rare for a machine to be installed without them. The same considerations discussed earlier about multi-degree-of-freedom behaviour apply in the isolation of machines as they do in the base-isolation of buildings, and the impedance of the supporting structure affects the actual isolation performance of anti-vibration mountings. As discussed in Chapter 2, characterising the vibration power of a source is complicated, and is best expressed in terms of the blocked force, being the force exerted on the machine mounting points, assuming it to be supported by an infinitely rigid foundation. Ways of measuring or determining the blocked force are outlined in Chapter 2.

The concern with the vibration isolation of static machines is two-fold. First, the key outcome is the reduction in vibration transmitted into the supporting structure, but second, it is necessary to consider the effect of resilient mounting on the motion of the machine itself to ensure that this does not become excessive. All connections between the machine and the unisolated structure have to be flexible to avoid short-circuiting the anti-vibration mountings. Even with the installation of flexible pipe connections, it must be borne in mind that at acoustic frequencies vibration capable of resulting in audible noise can be transmitted via the fluid in a connection pipe, and it may be necessary for pipes to be supported using resilient clamps or hangers.

It is frequently desirable to introduce an inertia block or other added mass below the machine and above its mountings in order to allow the use of less resilient mountings to achieve the required natural frequency. This is particularly necessary when a machine has several separate components such as a motor driving a pump, with the risk that introducing low stiffness mountings may cause unacceptable relative motion between the motor and the pump. A sub-frame or concrete inertia block helps to minimise the problems that would otherwise arise.

Moving sources – railways

While there are some similarities between the mathematical treatment of the vibration isolation of railways and of static machines, there are many major differences, caused not only by the fact that the source is moving, but also because the support system for the railway is a long linear system with mathematically complex characteristics, both as far as the propagation of waves along the track is concerned, and also the fact that there are frequently several layers in the system, including the rails, rail foot pads, sleepers or blocks. When a high level of isolation is required, railways may be installed on floating concrete slabs.

The source of excitation that gives rise to vibration from railways is primarily the wheel-rail interface. Neither the running surface of the rail head nor the wheel tread are ever perfectly smooth, and very small variations in the height of the rail along the band, where contact between the wheel tread and the rail head occurs, and in the radius of the wheel tread, where it makes contact with the rail, result in non-steady compression of the contact patch, being the small area of steel where the two surfaces meet. These dimensional variations are generally known as roughness and quantifying them is the subject of ISO 14837-34.[6] Due to compression of the steel, the contact patch is of finite dimensions – a few millimetres in width and length, and it has the properties of a spring. While in theory the spring constant depends on the load on the patch, as its size is load-dependent, it is normally possible to regard it as a simple linear spring. Irregularities of the running surfaces may take on greater magnitudes than can be described as roughness, as, for example, when a periodic undulation appears, known as corrugation. Where there have been wheel slides, spalling can occur, in which the crystalline properties of the steel change so that martensite is formed and pieces of the wheel tread break off (less common in modern systems with wheel-slide protection). The rail is not always continuous and may contain joints, either merely to connect lengths of rail, or to provide electrical insulation for signalling purposes.

Consequently, a primary method of mitigation of groundborne noise generation, when either roughness, if excessive, or other avoidable discontinuities are present, is the grinding of the rail head, and/or turning of the wheel tread on a lathe. Furthermore, if rail joints are present that are not essential, replacement of the rail with continuous welded rail (CWR) is beneficial. However, welding of joints in-situ does not necessarily achieve the same as replacement with CWR since some weld types involve the introduction of a metal alloy that does not have the same hardness as steel, and impulses may occur even though there is no visible rail joint.

The force that is created when the contact spring (usually called the Hertzian spring) is compressed and decompressed because of wheel/rail irregularities is dependent on the unsprung mass of the wheels. Unsprung mass is the mass of all components not supported by the primary suspension (usually a set of springs at the axle boxes), and in the simple case is the wheels, and their axles, if present, but may also include part of the load of the gearbox/traction system in motored bogies. Reducing the unsprung mass is not normally an option as a mitigation measure, but changes in rolling stock involving changes in unsprung mass are matters to take into account when considering differences in vibration generation between one case and another. It must be borne in mind that, while reduced unsprung mass results in reduced force, it also increases the natural frequency of the track and that reduces the vibration isolation effect of the track as a mass-spring vibration isolation system.

Even in simple cases, such as railways supported by baseplates on sleepers and ballast, the dynamic behaviour of the railway is far from straightforward as the propagation of bending waves along the rails takes place with frequency-dependent wave speeds, as explained in Chapter 5. At some frequencies vibration does not propagate at all. When resilient elements are incorporated in the track support system, the effect has to be considered, not simply in terms of inserting a sequence of springs under a line of masses, but also in terms of the effect on the propagation of vibration along the track. As well as the dynamic behaviour of the track itself, the dynamics of the supporting structure or formation beneath the track also have to be taken into account.

The need for vibration isolation of railways frequently arises in the case of railways in tunnels, which means that the dynamic behaviour of the tunnel and its lining is part of the system. When trains run at high speeds on the ground surface, train speeds may reach or exceed the phase speed or group velocity of surface waves, resulting in important consequences.

The design of track for vibration isolation may differ significantly in terms of the number of stages, that is, single stage involving merely resilient rail baseplates, or multistage involving rails resiliently supported from blocks, sleepers or slabs, in turn resiliently supported from the tunnel invert or ground formation. Slabs may be of short or very long length. Suppliers of proprietary systems or components use several ways of expressing their performance, either in terms of dynamic stiffness and loss factor of components, or insertion loss or insertion gain (insertion loss with opposite sign), obtained by subtracting a spectrum with the system in place from a spectrum for a reference system. The reference system may be described as "ballasted track" or may be a hypothetical extremely stiff track support system. The spectra concerned may be calculated using simple transmissibility equations involving lumped parameters or may be more thoroughly computed spectra of lower spectral density for a specific length of track.

Types of mitigating track support

The generation and transmission of groundborne noise and vibration from a railway source may be mitigated by a variety of systems:

Ballasted track:

- Ballast with soft rail pads
- Ballast with sleeper soffit pads
- Ballast with booted sleepers
- Ballast with under-ballast mat
- Floating ballast trough
- Ground treatment beneath the ballast.

Non-ballasted track:

- Embedded rail
- Resilient rail pads
- Resilient rail baseplates
- Resiliently booted blocks
- Resiliently booted sleepers
- "Minislabs" (short lengths of floating slab usually supporting two or three rail fastenings or sleepers/blocks).
- Floating slab track (continuous support)
- Floating slab track (discrete support).

The underlying principle involved in the use of mitigated track support is modifying the mobility of the track, including lowering of the natural frequency.

Dynamic behaviour of resiliently supported railway track

A rail forming part of a railway track is in the simplest case a beam on a resilient foundation. The basic equations that apply are set out in Chapter 5, where it is explained that the phase speed c_B of a bending wave is $c_B = \sqrt[4]{\frac{B}{\rho' - \frac{s'}{\omega^2}}}\sqrt{\omega}$.

The driving point impedance of a beam is given by $Z = 2\rho' c_B(1 + i)$. In a complex quantity, which is a ratio of two physical magnitudes. the physical meaning of $(1 + i)$ is that there is a phase lag between the applied force and the resulting velocity, and $(1 + i)$ can be expressed as a magnitude (modulus) of $\sqrt{2}$, and a phase angle (argument) of $\tan^{-1}(1) = \omega/4$.

It follows from this that the wavenumber $k(\omega) = \omega/c_B$ is $\sqrt[4]{\frac{\omega^2\rho' - s'}{B}}$. A starting point is the track deflection under the load of a wheel or an axle, and, under that condition, the excitation frequency is zero. Setting $\omega = 0$ gives a value of k that is independent of frequency $k(0) = \sqrt[4]{\frac{-s'}{B}}$ $= \frac{1}{\sqrt{2}}\sqrt[4]{\frac{s'}{B}}(1 - i)$. The rail deflection u as a function of distance from the load point is $u = Ae^{-ikx} = Ae^{-x/L + ix/L}$ where $L = \sqrt[4]{\frac{4B}{s'}}$. Using Euler's equation this can be written as $u = Ae^{-x/L}[\cos(x/L) + i\sin(x/L)]$. Taking into account the phase lag of $\omega/4$, the real value of u becomes $\sqrt{2}Ae^{-x/L}\cos(x/L + \omega/4)$. This may be written as $u = Ae^{-x/L}[\cos(x/L) + \sin(x/L)]$. It remains to determine the magnitude of A. Recalling that the governing equation is $B\frac{\partial^4 u}{\partial x^4} + su + \rho'\frac{\partial^2 u}{\partial t^2} = 0$, in the static case so $\frac{\partial^2 u}{\partial t^2} = 0$ the equation becomes $B\frac{\partial^4 u}{\partial x^4} + su = -F'$. F' must be in units of force per unit length, and in the case of a wheel supported on a rail, this becomes wheel load in Newtons divided by the length of the contact patch. The calculation can be simplified by letting the contact patch length tend to zero, so that the force per unit length becomes a Dirac function $\delta(0)$. The next step is to integrate the wheel load over the length of the contact, that is, integrating the bending wave equation with respect to ∂x, and evaluating the definite integral over the length of the contact patch centred on $x = 0$. The result is $B\frac{\partial^3 u}{\partial x^3} + s'ux =$ $4\left\{Ae^{-x/L}[\cos(x/L) + \sin(x/L))\right\}/L^3 + s'ux$ evaluated over the range x = $-\partial x/2$ to $\partial x/2$, which gives the result $8B/L^3$. Thus, $A = QL^3/8B$, so the rail deflection at $x = 0$ is $QL^3/8Be^{-x/L}[\cos(x/L) + \sin(x/L)]$ or $Q/2s'Le^{-x/L}[\cos(x/L) + \sin(x/L)]$. At the centre of the contact point where $x = 0$ this gives a rail stiffness of $2s'L$ in units of Newtons per metre.

This equation (known as the Zimmermann[7] equation, after the author who developed Winkler's paper[8] on the behaviour of a beam on a resilient foundation) is widely used in the analysis of the dynamic behaviour of point-loaded rail on a resilient support system. It is extended for the case of a sequence of loads by superposition of the rail deflection from each load (wheel) along the rail. It has limitations in that the track foundation is modelled as a series of independent, linear elastic springs, which is a simplification of the behaviour of real track systems. The Winkler model assumes that deflection at any one point depends only on the pressure at that specific point, with no consideration for the shear forces or coupling effects between adjacent springs/sections of the soil continuum, and the model assumes a constant, linear spring constant ("track modulus") for all track components.

If frequency is reintroduced into the equation by replacing s' by $s' - \omega^2 p'$, the loaded track natural frequency can be found as the frequency at which $\omega^2 m = 2s'L$ in which m is the unsprung mass of the wheel in kilogrammes. Given that L is a function of frequency, the solution is most readily found numerically (for example using the Excel "solver" add-in). Damping in the resilient rail support can be included by letting s' be a complex number. This quasi-static result is then used to estimate force transmissibility of the resiliently supported track taking an approach analogous to that of the single-degree-of-freedom mass-on-a-spring case.

The quasi-static approach is only a practical approximation at frequencies below the rail cut-on frequency, and at low train speeds. Once the transition is made from a static sequence of point loads to a moving train, many more considerations have to be taken into account. In the moving train case, bending waves propagate along the track, and when they reach an axle load there will be an impedance mismatch and partial reflection will occur. This will result in a backward-travelling wave which will encounter the preceding axle, so that a beam mode will be set up. Along a moving train there is a set of beam modes which have similarities to "pinned-pinned" modes in stationary beams, with wavelengths determined by the inter-axle spacing, the bogie pitch and the car length. In the moving train case, support components under the rail also exhibit more dynamic characteristics, and where, for example, there are booted blocks, they will tend to exhibit all the modes of a vibrating body in all the six degrees of freedom. If instead of blocks there are sleepers, these will exhibit transverse modes as well as rolling. Where there is a long floating slab track, wave motion along the slab will occur with different speeds relative to the rail. With short floating slabs, these, like blocks, are free to vibrate in all six degrees of freedom. When modes of these kinds occur at the same time as closely-spaced frequencies, enhanced dynamic effects can occur, in some cases contributing to the formation of periodic rail wear known as corrugation. These effects do not appear in simplified predictions of the insertion gain of a vibration-isolating track form. They can, however, be simulated in numerical modelling.

Insertion gain plots shown relative to another track support system such as ballasted track may be unhelpful, since they show a large gain where the loaded track natural frequency was in the reference case, and a reduced gain at higher frequencies where the reference track was providing some vibration isolation.

Clearly when the track performance is to be evaluated, non-track parameters need to be held constant, including the rail vehicle parameters (wheels, axles, bogies including primary suspension, secondary suspension and vehicle bodies), the rail, the foundation, which in the case of underground railways includes the tunnel lining, tunnel invert (those parts of it which are not altered by the insertion of the track support system being evaluated), surrounding soil and/or rock and soil/rock depth, and other features between the source and receptor.

Insertion gain is dependent on the choice of receptor. Railway tracks, and/or railway tunnels where relevant, are, of course, long, and when they act as radiators of vibration to the wayside or to buildings on or near the ground surface, the effective source may not only also be long (dependent on the track decay rate at the frequencies concerned), but may also result in the radiation of a line of vibration signals with progressively varying phase angle, such that radiation efficiency has to be determined and wave cancellation taken into account. There exist both a nearfield close to the track, and a far field that may be at the distance of the receptor. This results in a complex distance attenuation effect that may affect the way insertion gain figures can validly be used to predict mitigation performance at the receptor. Insertion gain immediately beneath the rail support will not necessarily be the same as insertion gain at the receptor.

While the insertion gain of a track support system can be calculated using the full FDTD model by comparing the output for the case in question with a hypothetical case involving a "very stiff" track, it is desirable to simplify the method to remove parameters which cannot be influenced by the track supplier (such as ground conditions or the tunnel design) and to create simplified code for rapid evaluation of predicted track performance.

Previous models have employed 2.5D or 3D numerical models[9], or have[10,11] approximated the track with beam, spring and mass elements coupled to an elastic half-space, or[12] with the track system represented by its impedance per unit length of track. Software for the calculation of vibration from surface and underground railways has been developed[13] that is capable of computing vibration-level differences for different track systems, assuming rail pads as continuous spring elements and sleepers as a continuous mass. The method described here models track support as discrete pads, baseplates, blocks or sleepers, as appropriate, using a simplified finite-difference algorithm, and calculates the insertion gain from the difference in the dynamic forces below the track supports.

Definition of insertion gain

In the railway context, the insertion gain of a track support system is defined, for example, in DIN 45673-4, as the difference in decibels, between the transmitted vibration forces with the track components relative to those without the component. It may be expressed as a spectrum (e.g., in 1/3 octaves). While this may refer to an actual track replacement, it may also be used to express the predicted vibration that occurs with a one-track system in comparison with a hypothetical base case.

Principles for calculating an insertion gain

A rail vehicle travelling on a track is a multi-degree-of-freedom dynamic system, the input signal to which occurs at the contact patch between the wheel and the rail. Looking upwards from the contact patch there is, in simple terms, a spring-mass-spring-mass-spring-mass system (if low frequency bending modes for the vehicle body are ignored). These springs and masses are: the Hertzian contact spring, the wheel and bogie unsprung mass, the primary suspension, the bogie sprung mass, the secondary suspension and the vehicle body. Looking downwards from the contact patch there is a complex mass-spring-mass-spring-mass-spring-resistance in which the components are the rail, the rail foot pad, the sleeper or block, the sleeper support (ballast or a resilient boot/pad) and the foundation in the form of tunnel invert/lining/soil/rock. In a mass-spring-mass-spring system, the result is two coupled natural frequencies, one of which is higher than the higher of the individual mass-spring frequencies, and the other is lower than the lower of the individual mass-spring frequencies.

An important example of what happens when a sequence of masses and springs are coupled together is the behaviour of the Hertzian contact spring, which is further discussed in the next section.

Key parameters

The key parameters that affect the insertion gain of a track system include the geometry of the track system (such as fastener spacing), the mass and bending stiffness of the rails, the mass of the sleepers (if present) and the acoustic stiffness of the rail or sleeper supports.

Other key parameters will affect the insertion gain, such as the mass and suspension of the rolling stock, the Hertzian contact stiffness at the rail-wheel interface, the tunnel and the ground below the track. These are parameters that the track-system supplier will have no control over. When comparing track systems, these parameters should be kept constant, or their effects removed from the insertion gain.

Methodology

The methodology starts with a process of simplification of the full 3D model in order to observe the effect associated with a series of simplifications. These include (1) representing the track as a single rail, (2), replacing the line of axles with a single axle, (3) omitting the bogies and vehicle bodies, (4) removing the influence of the lithology (5) removing the influence of the tunnel lining and (6) assuming frequency-independent damping. These simplifications are identified and illustrated in Figure 11.7, which shows the insertion gain calculated using the full source-path-receiver model compared with the simplified models.

A difficulty encountered with the use of a fully detailed model of the system arises when modelling the "very stiff" track support system as a reference case. The effective rail mass, M_r, in a resiliently supported system is much less than the wheel mass, and the natural frequency, f_o, associated with the rail-contact-wheel system is close to the mass-spring natural frequency of the rail and the Hertzian spring, $f_o = 1/2\pi\sqrt{(k_h/M_r)}$ where k_h is the Hertzian spring

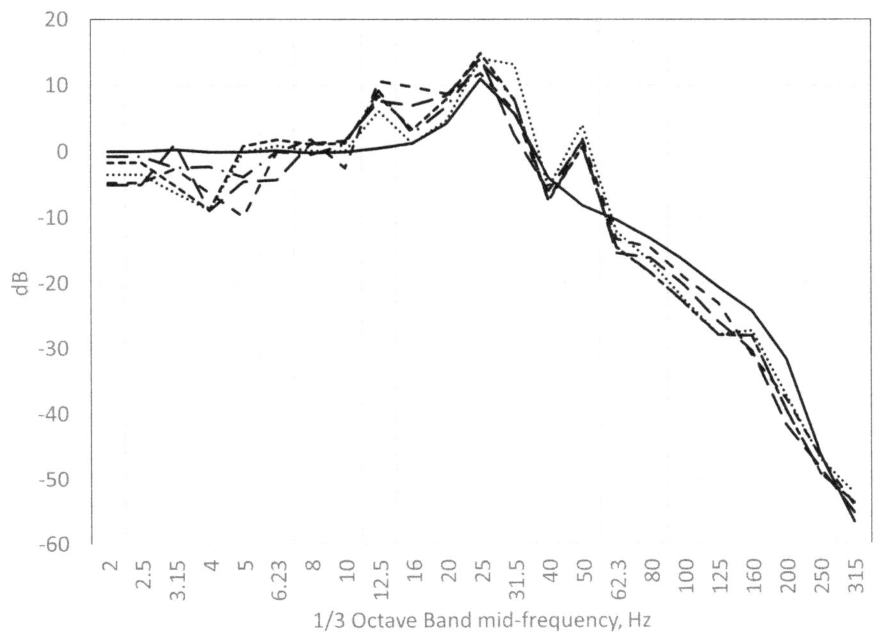

Figure 11.7 The effect of simplification. Key: —— full model; - - one rail; — · — one rail, one axle; - - one rail, one axle, unsprung mass only; ···· one rail, one wheel, unsprung mass only, frequency-independent damping; —— one rail, one wheel, unsprung mass only, 1D model.

stiffness, and is typically over 700 Hz. This is above the frequency range of interest. In the case of very stiff rail support, the natural frequency becomes close to that of the wheel and the Hertzian spring, $f_o = 1/2\pi\sqrt{(k_h/M_w)}$ where M_w is the mass of the wheel, and is well within the frequency range of interest, typically about 180 Hz. Consequently, the spectrum of vibration transmitted into the foundation by a train on a very stiff track will have a prominent peak within the frequency range of interest and reduced levels of vibration above that frequency. This peak gives a false indication of the performance of the track support, since it is not attributable to the behaviour of the track support system in a relevant way. The means of overcoming this problem, which occurs due to the choice of a reference system that is effectively rigid and does not represent any actual real-world track support system, is to raise the stiffness of the contact patch in the rigid system to a hypothetical value many times greater than its true value, so that the contact-wheel natural frequency is raised to a value above the frequency range of interest. (The stiffness has to remain finite in order to furnish the required input force through compression of the Hertzian spring).

Having simplified the model in this way, it is also possible to simplify the finite-difference code to fewer than 500 lines which could be ported to any convenient language. The approach taken is to apply a relative displacement between the wheel and the rail and, thus, to compress the Hertzian contact spring at the wheel/rail contact patch, and to integrate the resulting force at the base of all the lowest resilient elements along the track, which is extended in length so as to be beyond the 10 dB down-points for the track decay rate. The relaxation damping algorithm used in the full 3D model can be retained in order to be able to model frequency-dependent (or frequency-independent) damping.

The simplified finite-difference code

The finite difference method is employed in order to solve the bending wave equation for an Euler beam

$$EI\,\partial^4 u/\partial x^4 = -m'\partial^2 u/\partial t^2 \tag{11.3}$$

where E is the Young's modulus of the beam and I is the second moment of area of the section; m' is the mass per unit length; u is the vertical displacement; x is distance along the beam and t is time. This equation neglects rotary inertia and shear, which have an effect on the phase velocity of bending waves in a beam of less than 10% if the wavelength is greater than $6h$ where h is the height of the beam.

The finite-difference approximation of Equation (11.3) is obtained by dividing the rail into 2000 elements approximately 200 mm in length (the exact value depending on the fastener spacing to be studied) and for each element of the rail iterating the following two equations

$$\begin{aligned} ddx4 = {} & (((u[i+2]-u[i+1])-(u[i+1]-u[i]))- \\ & ((u[i+1]-u[i])-(u[i]-u[i-1])))- \\ & (((u[i+1]-u[i])-(u[i]-u[i-1]))- \\ & ((u[i]-u[i-1])-(u[i-1]-u[i-2]))) \end{aligned} \tag{11.4}$$

$$v[i] = v[i] - ddx4^{*}EIdx4 \tag{11.5}$$

where i is the number of the rail element; EIdx4 is the product of the bending stiffness EI and the length of the rail element dx raised to the fourth power; $u[i]$ is the vertical displacement of the rail and $v[i]$ is the vertical velocity.

In the model described here, it is assumed that the rail is supported under every third element by a pad of stiffness (kpad), below which is a block of mass (mblock). Each block in turn is supported by a resilient boot of stiffness (kboot).

Following the execution of code implementing Equations (11.4) and (11.5) or each element i, the force exerted by the deflected rail pad at every third element is computed in terms of the incremental velocity as

$$\mathrm{dvr} = -(u[i] - \mathrm{ub}[i])*\mathrm{kpad}/\mathrm{mrail}/\mathrm{dx}^{*}\mathrm{dt} \tag{11.6}$$

$$\mathrm{dvb} = ((u[i] - ub[i])*\mathrm{kpad} - ub[i]*\mathrm{kblock})/\mathrm{mblock}^{*}\mathrm{dt} \tag{11.7}$$

where $ub[i]$ is the vertical displacement of the block, dvr is the incremental velocity of the rail and dvb is the incremental velocity of the block and dt is the finite difference time step.

Above the rail is the wheel, treated as a lumped mass on the Hertzian contact spring, with the lines

$$vw = vw - (uw - u[i])*\text{kpatch}/\text{mwheel}^*\text{dt}$$
$$v[i] = v[i] + (uw - u[i])*\text{kpatch}/\text{mrail}/\text{dx}^*\text{dt}$$
$$uw = uw + vw^*\text{dt} \tag{11.8}$$

where 'kpatch' is the stiffness of the Hertzian contact patch.

Damping is modelled using Boltzmann's relaxation model[14]

$$\sigma(t) = D_1\epsilon(t) - \int_0^\infty \epsilon(t - \Delta t)\phi(\Delta t)d(\Delta t) \tag{11.9}$$

where $\sigma(t)$ is the stress at time t, $\epsilon(t)$ is strain, $\phi(\Delta t) = \frac{D_2}{\tau}e^{-\Delta t/\tau}$, D_1 and D_2 are constants and τ is a relaxation time.

Equation (11.9) may be replaced by a sum of three terms with different relaxation times, and by choice of values of D_1, D_2 and τ any desired frequency-dependence of the resulting damping effect can be achieved.

To implement Equation (11.9) the lines in the code for the rail are

$$
\begin{aligned}
\text{de} &= -(a[i] - ub[i])*\text{kdamp}/\text{mrail}/\text{dx}^*\text{dt}\\
\text{sp}[i][0] &= \text{sp}[i][0]*\text{phi0} + \text{de}^*\text{dt};\\
\text{sp}[i][1] &= \text{sp}[i][1]*\text{phi1} + \text{de}^*\text{dt};\\
\text{sp}[i][2] &= \text{sp}[i][2]*\text{phi2} + \text{de}^*\text{dt};\\
\text{dv} &= \text{dv} - \text{sp}[i][0]*\text{D2}/\text{tau0} - \text{sp}[i][1]*\text{D2}/\text{tau1} - \text{sp}[i][2]*\text{D2}/\text{tau2}\\
v[i] &= v[i] + \text{dv}
\end{aligned}
\tag{11.10}
$$

with corresponding lines for the block.

The use of a length of rail of the order of 400 m means that the ends of the rail may be connected end-to-end to avoid reflections and maintain stability, and the length of the modelled rail is long enough for the bending waves to have decayed to insignificant amplitudes at the ends. The time step used is 1/131072 seconds, and the routine is run for 262144 iterations, that is, 2 s. The run time is only a few seconds.

Excitation of the system is achieved by an initial condition at $t = 0$ consisting of a finite displacement of the Hertzian contact spring, that is, a Heaviside step.

Output takes the form of the force under the boot below each block, as a time series. Each of these series is then subjected to FFT. As a result of the simplifications described above, the phase relationships between the forces below each block may differ from those in the full 3D case. To discard phase effects, the squares of the forces at each block are summed to give a power-related total output, as seen by the rigid support below the boots along the full length of the rail, expressed in decibels. This is subtracted from the corresponding curve for the rigid system to give the insertion gain (IG) in dB. The curve for the rigid system is found from the IG, at frequencies well below the loaded track natural frequency, and has a straight-line slope of −9 dB per octave.

The results take the form of 1/3 octave band spectra of IG, which show the eigenmodes of a coupled two-stage system, taking account of the effect of damping, and the slopes to the insertion gain curves between and above the frequencies of the eigenmodes. For a single-stage system, such as slab track with periodic rail supports, the code omits the lines for the blocks and boots. Examples are given in Figure 11.8. In the results for two-stage systems, the characteristic loaded track resonance is clearly evident. The coupled natural frequency of the blocks between the rail pads and the resilient boots also appears, reducing the insertion gain around 160 Hz–250 Hz in the examples shown in Figure 11.8. This effect is absent in the case of the single-stage slab track, but the enhanced attenuation that occurs above the coupled frequency in the two-stage systems is also absent.

The properties assigned to each of the cases in Figure 11.8 are as given in Table 11.1.

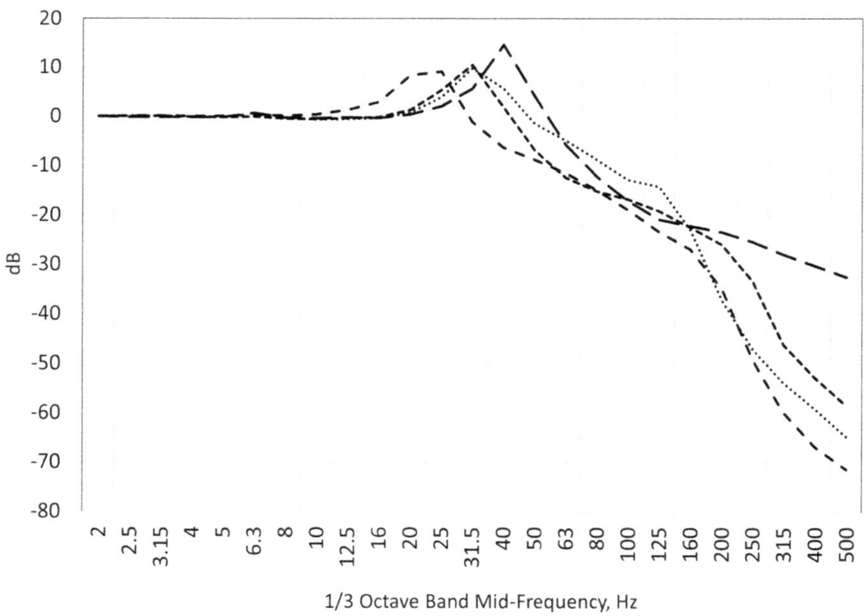

Figure 11.8 Simplified IG of track support systems, dB re rigid track. Key: — — slab track; - - booted block low mass; – – booted block low stiffness, high mass; ···· booted sleeper.

Table 11.1 Input parameters for results in Figure 11.8

Track type	*Rail pad stiffness MN/m*	*Rail pad damping (dimensionless)*	*Block mass kg*	*Boot stiffness MN/m*	*Boot damping (dimensionless)*	*Rail mass kg/m*	*Fastener spacing, m*	*Wheel mass, kg*
Slab track	40	0.15	-	-	-	60	0.65	950
Booted sleeper	150	0.2	210	31.5	0.2	60	0.75	950
Booted block low mass	100	0.2	45	28	0.2	60	0.6	950
Booted block high mass	200	0.2	120	14	0.2	60	0.6	950

MITIGATION BY TRANSMISSION PATH INTERVENTIONS

Through analogy with airborne acoustics, the question arises whether an equivalent to a noise barrier can be inserted in the ground to attenuate the transmission of groundborne noise. The analogy is only partial, however, because, unlike the airborne case, there are several different wave types involved, and whereas above a noise barrier there is usually nothing to cause downward reflections (although temperature and wind gradients have counteracting effects). In the case of groundborne noise, two options are theoretically possible: (1) a trench and (2) a barrier such as a line of sheet piles. Completely open trenches are somewhat impracticable because of the difficulty of keeping them open, and a low stiffness filler may be considered.

Trenches

The topic of vibration attenuation by the use of trenches was reviewed By Mahdavisefat et al.[15] Their conclusions were that trench depth was the most important with a depth 1.5 λ_R being the recommended minimum. Trench width was found to be less important, and some researchers found that excessive normalized width results in vibration amplification. The source-barrier distance was found to have a low effect on barrier performance. Of all material properties that influence trench performance, the ratio of shear-wave velocity in filling material to that of the surrounding soil has the most significant role. The use of a material with high-energy absorption and damping capabilities results in greater amplitude reduction of the input waves and a more reliable level of isolation.

Multiple trenches have been found to be very effective and require considerably lower depth values to provide the same amplitude reduction compared to single trench systems. Trenches that employ improved non-rectangular, cross-sectional forms perform better than their rectangular counterparts and can have stability advantages.

Thompson et al.[16] and Jiang et al.[17] carried out a combined finite-element analysis/boundary element method (FEA/BEM) study which concluded that, in all the ground conditions considered, an open rectangular trench performs best. The depth is a more important parameter than the width that has little influence on its performance. Barriers containing a soft-fill material were shown to be much less effective than an open trench, but still have some potential benefit. It was found that the stiffness of the barrier material and not its impedance is the most important material parameter. For the open trench 6 m deep, a reduction of the order of 20 dB was modelled at frequencies above 10 Hz. However, an open trench of that depth would not be stable and with more practical materials filling the trench the reduction was about 5–10 dB. Similar results were obtained by Auersch.[18]

In cases where there is layered ground, the effect of a trench is limited to the difference between the amplitude of the surface wave that is interrupted by the trench, and the amplitude of waves reflected from the layers with significant impedance mismatches. This can be discovered in advance of trench installation by recording the arrival of waves from an impulse source (such as an accelerated hammer blow) near to the vibration source. The time-domain signal may show distinct arrival times for the surface wave and the reflected waves, from which it is possible to extract their relative amplitudes.

Where there is not a significant contribution from reflected waves from subterranean layers, the principal effect of a trench is to interrupt the passage of Rayleigh waves. The amplitude of Rayleigh waves is depth dependent and frequency dependent to the extent that longer wavelength components of the signal penetrate deeper into the ground, as shown by Figures 3.5 and 3.6 in Chapter 3. This makes the "noise barrier" effect of a trench more frequency dependent than is the case for an airborne sound barrier, where frequency-dependence is an effect of diffraction of sound over the top of the barrier. Diffraction under the base of a trench also occurs, but, in addition to that effect, the low frequency components of the signal are propagated at greater depth.

Numerical methods of predicting trench performance are well-suited to the problem of predicting the performance of vibration-reducing trenches.

The inverse of the trench is the insertion of a rigid barrier, such as a line of sheet piles. This approach was investigated as part of the RIVAS programme.[19, 20] A sheet pile wall 100 m in length, 12 m deep, with every fourth pile extended to 18 m was installed 5.60 m from centre of nearest track. The conclusions were, for the passenger trains, that the sheet pile wall resulted in vibration reduction from 4 Hz up to a frequency of 16 Hz which generally increased with frequency. Negative insertion loss values around 31–40 Hz and 80 Hz were measured, but were not observed in any predictions made for stiff wave barriers and were, therefore, believed to have other causes. For freight trains, high insertion loss values of up to 5–10 dB were obtained in a broad frequency range. The performance generally decreased with increasing distance behind the sheet piling wall.

The RIVAS study also contains the results of studies on a soft wave barrier, a jet grouted wall and heavy masses along the track.

For the soft wave barrier, a 3 m deep barrier of 50 mm thick reconstituted rubber (of trade name Regupol) was inserted, and an insertion loss of about 10 dB from 50 Hz upwards was found.

For the jet-grouted wall, the findings were that, immediately behind the wall, insertion loss values of about 5 dB were obtained from 8 Hz upwards, with a (measured) peak of about 12 dB at 25 Hz, which was also the frequency range where the highest vibration levels were found during a train pass-by. The performance decreased with increasing distance from the track. A slight amplification of vibration levels was observed just in front of the jet-grouted wall (at 14 m).

The study of heavy masses along the track was numerical and related to a line of gabions (wire baskets filled with rocks and stones). The simulated effect was that the wall started to be effective above the mass-spring resonance frequency around 10 Hz. For all distances from the track, a reduction up to 5 dB was seen around 16–20 Hz. At higher frequencies; the effect strongly decreased with increasing distance from the track.

Where linear mitigation measures are installed, an important consideration is the length of the installation, because it is only effective with respect to parts of a train that are behind it from the point of view of the receiving location. As a train approaches a line of mitigation, the groundborne noise level will fall in level until the whole train is shielded and then rise again as the train emerges at the other end. This may cause increased annoyance from vibration and groundborne noise compared with the unmitigated case.

BOX-IN-BOX

In addition to reduction of groundborne noise in transmission, it is possible to reduce it at the receiver by installing a resilient lining to the receiving room, often referred to as "box-in-box" construction. This approach involves the installation of a completely separate, resiliently supported inner skin to a room, in which there are no solid bridges across the separation between the basic structure and the resilient linings. It may involve a floor supported on resilient bearings, on which is constructed an independent wall and ceiling system, or a resilient floor and resiliently attached wall linings and a resiliently suspended ceiling. All services entering the room, such as pipes and ducts as well as electrical connections must have flexible connections in them.

Heckl[21] provided an equation for calculating the marginal improvement ΔR in the sound reduction of a stiff wall with the addition of a resilient skin. The same principles also apply to a floating floor and suspended ceiling. For point support of the resilient layer the equation is

$$\Delta R_{point} = -10\log\left[\beta^2\frac{8}{\pi^3}n''\frac{c^2}{f_c^2} + \left(\frac{f_0}{f}\right)^4\right] + 10\log\sigma_{rad_1} \tag{11.11}$$

where β is the vibration isolation of the flexible support; n'' is the number of attachment points per square metre; c is the speed of sound in air; f_c is the critical frequency of the resiliently supported layer $\frac{C^2}{2\pi}\sqrt{\frac{\rho_S}{B}}$; $B = Et^3/12$; E is Young's modulus of the resilient layer and t is its thickness; f_o is the resonant frequency given by $\frac{1}{2\pi}\sqrt{\frac{s}{\rho_s}}$ where s is the combined stiffness of the panel's supports and the airspace between the structural wall and the resiliently supported lining; ρ_s is the mass per square meter of the lining and σ_{rad1} is the radiation ratio of the structural wall. This equation is only valid in the region $f < f_c$.

When the point support is a resilient fastener, β will be less than 1.0. Its value can be approximated as $1/(1 - \omega^2 m/k)$ where m is the mass per square metre of the lining and k is the stiffness per square metre of the point supports. In the region of the natural frequency $\sqrt{(k/m)}/2\omega$ the added resilient layer will diminish the amount of sound reduction, by an amount depending on the damping in the resilient supports. Above $\sqrt{2}$ times the natural frequency, there will be additional sound reduction. The maximum sound reduction depends on the relationship between the critical frequency of the structural wall, and that of the added resilient layer. It will also be limited by flanking transmission at the edges of the wall unless the resilient layer is fully isolated around its edges and the medium used for isolation has a combination of mass and low stiffness.

Heckl also gives an equation for line support instead of point support

$$\Delta R_{line} = -10 \log \left[0.64 n' \frac{c}{f_c} + \left(\frac{f_0}{f} \right)^4 \right] + 10 \log \sigma_{rad_1} \qquad (11.12)$$

It is important to take account of the fact that often groundborne noise occurs at relatively low frequencies, in which case the conventional room acoustics equations for diffuse fields in rooms do not apply. Instead, it is necessary to use a modal acoustics approach, since a rectangular room with dimensions comparable to the wavelengths involved will respond at the frequencies of room modes, and these are not as numerous at low frequencies as they are higher up the spectrum, where statistical energy approaches can be applied, and room spaces can be assumed to be diffuse.

When a resilient layer is attached to a wall, floor or ceiling, not only does Equation (11.11) have to be considered, but also the resulting change in radiation performance of the innermost surface. In a diffuse field, this would be a matter of radiation ratio, but in modal acoustics, the coupling between the surfaces of a room and the air in the room must be evaluated in terms of the relationship between panel and room modes. When a resilient lining is attached to a room surface, not only will the vibration amplitude change, but also the shape of the plate modes may also change, along with the frequencies at which they occur. This may be particularly marked if the plates which form the boundaries of the unlined room behave as if they have clamped edges, and the installed resiliently supported lining behaves as if its edges are simply supported. The equations in Chapter 5 define the difference between plate modes for plates with simple support and clamped edges. While Table 5.1 of Chapter 5 shows the difference for a concrete plate 200 mm thick, it is useful to compare a clamped 200 mm concrete plate with a resiliently attached lining panel of 50 mm plasterboard assumed to have simple edge support, as shown in Table 11.2. All but two of the plasterboard modes are at frequencies below the lowest mode of the concrete plate.

Table 11.2 Comparison between plate modes of 200 mm concrete and 50 mm plasterboard for plate 2.5 m by 4 m

m	*n* 1	2	3	4	5
1	139	346	658	1075	1595
2	175	351	652	1063	1580
3	276	411	686	1085	1595
4	428	526	761	1133	1628
5	626	696	887	1221	1691

Clamped 200 mm concrete

m	*n* 1	2	3	4	5
1	30	53	77	102	127
2	40	59	82	105	130
3	53	69	89	111	134
4	68	81	98	119	141
5	83	93	109	128	148

Simply supported 50 mm plasterboard

When a plate, such as a floor slab, has a resiliently supported layer attached to it, the impedance of the resilient layer affects the dynamic behaviour of the supported layer. At the mass-spring natural frequency of the added system, the bending wave speed in the support layer tends to infinity, which means that all of the supported panel moves vertically in-phase, the lowest plate mode in the supported panel is increased and the equation for the panel modes of the resiliently supported layer becomes

$$\omega_{(m,n)} = \sqrt{\frac{B}{m''}\left[\left(\frac{n\pi}{l_x}\right)^2 + \left(\frac{m\pi}{l_y}\right)^2\right]^2 + \frac{k''}{m''}} \qquad (11.13)$$

where $B = \frac{Et^3}{12(1-\sigma^2)}$; E is Young's modulus; t is the panel thickness and σ is Poisson's ratio, m'' is the mass per unit area of the panel and k'' is the stiffness per unit area of the resiliently supporting layer; l_x and l_y are the panel dimensions and n and m are any integers. The stiffness of the supporting layer will be a complex number because of its inherent damping, and indeed damping will control the amplitude of peaks in the spectrum of panel modes.

In Chapter 8, Equation (8.6) gives the relationship between the velocity at a point on the surface of a panel forming a boundary to a rectangular room and the sound pressure level at a point in the room. This equation can be used to compare the relative sound radiation from a 200 mm concrete panel and a 50 mm plasterboard panel.

RAISED ACCESS FLOORS

Office developments are frequently constructed with raised access floors, where composite panels consisting, for example, of a layer of calcium sulphate board between two steel sheets, are supported by pedestals, typically on a 600 mm × 600 mm grid. In cases where groundborne noise enters the building and potentially radiates sound from the floor into the room above, it becomes necessary to evaluate the effect of the presence of the raised access floor on transmission of vibration from the concrete slab, through the pedestals and the raised panels and into the room above.

Some of the principles involved are the same as those for box-in-box construction. However, some of the relevant parameters are different. First, the mass of the floor panels tends to be relatively low as they are typically 50 mm or less in thickness. They are supported on a grid that creates a simply supported edge condition, and it is usually found that the lowest plate resonance frequency is several hundred Hertz, above the peak frequency in the groundborne noise. Furthermore, although the void beneath the floor panels may be quite deep, such as 250–350 mm, the mass-spring natural frequency of the panels on the airspace can also be above the peak frequency of the groundborne noise. For example, a layer of panels of 30 kg/m^2 surface density above a 300 mm airspace, neglecting the effect of the support pedestals, has a natural frequency of $\sqrt{(k/m''d)}/2\pi$ where k is the bulk modulus of air (1.4 kPa); m'' is the mass per square metre and d is the depth of the cavity in metres, giving the result of 20 Hz. Treating the system as a single-degree-of-freedom mass-on-a-spring tells us that below $\sqrt{2} \times 20 = 28$ Hz there will be amplification of vibration (peaking at 20 Hz), and that attenuation will occur above 28 Hz, increasing with frequency, dependent on the damping in the system. Damping of the air cavity will be provided by the services contained within it, which will also occupy part of the airspace, so that the spring rate of the airspace will actually be greater than k/d.

Normally, in the case of a secondary panel as an added layer above a vibrating surface its performance will be limited by the critical frequency of the panel (the frequency at which the bending wavelength in the panel is equal to the wavelength of sound in air at that frequency). However, if the panels are simply supported and only of relatively small dimensions, such as 600 mm × 600 mm, bending wave propagation will not occur in the same way as it would with a continuous panel. On calculating the plate modes of a square simply supported panel, taking into account the bending stiffness of the panel that is designed to be high enough to support the required point loads, it emerges that bending waves do not occur in the panel system at the frequencies of most groundborne noise. As a result, the radiation efficiency of the raised floor is not significantly different from that of the main floor slab. The effect of the pedestals, however, needs to be taken into account. The standard pedestal design results in point excitation of the floor at intervals determined by the grid size, such as 600 mm. In the case where the lowest plate mode of the floor plates is above the excitation frequency, i.e. that is, the vibration transmitted via the pedestal, and for a pedestal that has high stiffness in vertical compression, there is transmission from the slab into the raised floor without loss.

NOTES

1. Harris, C.M., *Handbook of Noise Control*, McGraw-Hill Book Company, New York, 1957.
2. Clough, R.W., and Penzien, J, *Dynamics of Structures*, McGraw-Hill Book Company, New York, 1975.
3. Bycroft, G.N., "Forced vibration of a rigid circular plate on a semi-infinite elastic space and on an elastic system." *Philosophical Transactions of the Royal Society*, 248, A.984, 1956.
4. Thornely-Taylor, R.M., "Underground acoustics." *Proceedings of the Institute of Acoustics*, 38(Pt.1), 1–19, 2016.
5. Thornely-Taylor, R M., "The use of numerical methods in the prediction of vibration." *Proceedings of the Institute of Acoustics*, 29 (Pt.1), 2007.
6. ISO/TS 14837-34:2024(en). "Mechanical vibration—Ground-borne noise and vibration arising from rail systems—Part 34: Characterizing irregularity of the running surfaces with respect to vibration excitation." 2024.
7. Zimmermann, H., *Die Berechnung des Eisenbahn Oberbaues*, Verlag Von Ernst & Korn, Berlin, 1888.
8. Winkler, E. Die Lehre von der Elastizität und Festigkeit, Theory of elasticity and strength of *materials*, Verlag von H. Dominicus, Prague, 1867.
9. He, C., Zhou, S., Guo, P., Di, H., Zhang, X., "Modelling of ground vibration from tunnels in a poro-elastic half-space using a 2.5D FE-BE formulation." *Tunnelling and Underground Space Technology*, 82, 211–221, 2018.
10. Burgemeister, K., and Greer, R. "Using insertion gains to evaluate railway vibration isolation systems." *Proceedings, Acoustics 2004*, Gold Coast, Australia, 3–5 November 2004.
11. Jones, C.J.C., "Groundborne noise from new railway tunnels (Invited Paper)." *Proceedings of Internoise*, Liverpool, UK, 30 July–2 August 1996, Book 1, 421–426.
12. Guigou-Carter, C, Villot, M., Guillerme, B., and Petit, C., "Analytical and experimental study of Sleeper SAT S 312 in slab track Sateba System." *Journal of Sound and Vibration*, 293, 878–887, 2006.
13. MOTIV Homepage: https://motivproject.co.uk
14. Boltzmann, L., "Zur Theorie der elastischen Nachwirkung" *Annalen der Physik*, 7, 624–654, 1876.
15. Mahdavisefat, E., Heshmati, A., Salehzadeh, H., Bahmani, H., and Sabermahani, M. "Vibration screening by trench barriers, a review." *Arabian Journal of Geosciences*, 10, 513, 2017.
16. Thompson, D., Jiang, J., Toward, M., Hussein, M., Ntotsios, E., Dickmans, A., Coulier,P., Lombaert, G., and Degrande, G. "Reducing railway-induced groundborne vibration by using open trenches and soft-filled barriers." *Soil Dynamics and Earthquake Engineering*, 88, 45–59.

17. Jiang, J., Toward, M., Dijkmans, A., Thompson, D., Toward, Lombaert, G., Degrande, G., and Ryue, J., "Reducing railway induced ground-borne vibration by using trenches and buried soft barriers." *Proceedings of the 11th International Workshop on Railway Noise IWRN11*, 2015.
18. Auersch, L., "Reduction of train-induced vibrations—calculations of different railway lines and mitigation measures in the transmission path." *Applied Science*, 13, 6706, 2023. 10.3390/app13116706
19. Coulier, P., Degrande, G., Dijckmans, A., Houbrechts, J., Lombaert, G., Rücker, W., Auersch, L., Rodríguez Plaza, M., Cuellar, V., Thompson, D., Ekblad, A., and Smekal, A. "WP4—Mitigation measures on transmission/propagation." EU FP 7 Project RIVAS (Railway Induced Vibration Abatement Solutions) D4.1-Report_defining_scope_and_constraints_of_numerical_parametric_study.pdf
20. Lombaert, G., Dijckmans, A., Coulier, P., Degrande, G., Smekal, A., Ekblad, A., Thompson, D., Jiang, J., Toward, M., Rücker, W., Keil, J., Vukotic, G., Rodríguez Plaza, M., Andrés, Á., Cuéllar, V. Garburg, R., and Müller R., "Design guide and technology assessment of the transmission mitigation measures deliverable." D4.6 Project RIVAS (Railway Induced Vibration Abatement Solutions) http://www.rivas-project.eu/fileadmin/documents/D4.6-Design_guide_and_technology_assessment_of_the_transmission_mitigation_measures.pdf
21. Heckl, M., "Untersuchungen uber die Luftschalldämmung von Doppelwänden mit Schallbrücken." *Proceedings of the 3rd International Congress on Acoustics*, Stuttgart, 1959.

Chapter 12

Uncertainty

Throughout the process of measuring, predicting, assessing or mitigating groundborne noise, it is necessary to consider uncertainty, in terms of the probability that the result obtained will differ from the "true" figure. While there are many causes of uncertainty in acoustics, in the field of groundborne noise there are many more, and the magnitude of potential uncertainty is greater.

Uncertainty may arise from lack of knowledge of the values of parameters, error or inability to calculate or model physical processes accurately enough or in sufficient detail.

UNCERTAINTY IN MEASUREMENT

Let us start with measurement, and list the potential causes of uncertainty:

- Transducer type and location
- Transducer fidelity
- Measurement/recording chain fidelity
- Signal processing algorithms.

Transducer type and location

The characteristics of transducers, in particular geophones and accelerometers are discussed in Chapter 9. Important aspects of these are the behaviour of geophones in the vicinity of their natural frequency, where it is necessary not only to compensate for the innate enhanced sensitivity of the transducer in that region, but also to be aware that the signal will undergo phase distortion. In many measurements, phase is of no particular interest, and as long as the frequency response of the transducer falls within the allowable range, phase can be ignored. The natural frequencies of accelerometers are well above the frequency range of interest in groundborne noise, and the principal issue is the adequacy of the accelerometer sensitivity to ensure that

 DOI: 10.1201/9781003618775-12

what is measured has a great enough signal-to-noise ratio, taking the self-noise of the entire measurement chain into account. Even if the signal-to-noise ratio is good, it is also necessary to ensure that the signal is not so low in the range of the analogue-to-digital converter (ADC) to give results in terms of only a few bits at the bottom of the scale, which will greatly reduce the resolution of the signal.

The method of mounting and fixing the transducer is the largest potential source of uncertainty, or in this case, error, in the results. The presence of the transducer can modify the behaviour of the surface, whose vibration it is to measure, and the method of attachment can modify the transmission of vibration from the surface to the transducer. As illustrated in Chapter 9, the potential error from this cause can be large. In addressing the problem and looking to minimise the error, it is worth rethinking the purpose of the measurement. The greatest difficulty arises when the quantity of interest is vibration in the ground below an open site, where the top surface may be vegetation, soil, made ground, remnants of previous structures, an exposed rockhead, or the top of a pile, as sometimes survey engineers request piles to be inserted specifically for the purpose of measuring vibration from an underground source such as a railway tunnel.

The actual vibration of the surface of an open site is seldom of direct interest. Such measurements are nearly always done because it is intended to construct a sensitive building on the site, or nearby to the site. It is important to look ahead to the use to which the results will be put, and in particular to the likely relationship between vibration on the open site and the proposed future structure. This necessitates taking into account the nature of the source and the types of waves that are being emitted from it.

Occasionally, the reason for measuring vibration on the surface of an open site may be as a surrogate for vibration in a nearby building to which access is not possible.

Although many sources of advice refer to burying the transducer as a way of installing it, that usually means it will be measuring vibration in a medium that is capable of being dug with a spade, and there are very few reasons why vibration in that kind of medium is of interest. It is often possible to remove the upper layer to reveal a reasonably competent compacted layer of soil even if this means having a trial pit dug as part of a geotechnical investigation. A DIN plate with accelerometers attached can then be bedded firmly on this surface. The same considerations are true of cast patches of plaster-of-Paris which are sometimes recommended, and of spikes.

What is important is the unsuitability of apparently convenient hard features such as concrete blocks in the ground remaining after demolition of previous structures on the site. Vibration in the block will depend on the characteristics (including dimensions and mass) of the block, which may be difficult to know, and to calculate or predict the relationship between vibration of the block and of the structure that will be there in the future is unnecessarily difficult.

The case of a pile inserted for the purpose of measurement of vibration from an underground railway is more complex than it appears. Such a pile will not behave in the same way as a future pile as part of the foundations of a building, which will have a pile cap on top connected to either ground beams or a concrete slab. Vibration on top of a pile of this kind will diverge from that on the ground around it by small amounts, and unless it is possible to carry out numerical modelling of the dynamic behaviour of the pile, its main effect will be to increase uncertainty rather than reduce it.

Transducer fidelity

The fidelity of the transducers will be obtainable from their calibration results. If a portable calibrator is used at the time of the measurements, the overall sensitivity of the transducer and the measurement/recording chain will be known, although nearly all calibrators operate at a single frequency so that only overall sensitivity at that frequency will be checked. The uncertainty associated with transducer fidelity is one of the smallest of the set of the causes of uncertainty in the measurement process.

Measurement/recording chain fidelity

The response of the measurement/recording/logging system is obtainable from the calibration results, supported by the use of a portable calibrator. This part of the process is more vulnerable to error than measurement uncertainty if the system relies on user input of sensitivities and range settings. Recording a calibration signal is a good way of checking that sensitivities and range settings have been correctly set and noted.

Signal processing algorithms

Some systems have built-in processing facilities to provide output in terms of the measurement units of interest. Others may simply log the time-domain signal for later analysis using computer software. The processing required will consist of frequency-weighting, if it is required to use the measurement results for the purpose of estimating likely levels of re-radiated groundborne noise in dB(A), and exponential averaging using the "F" or "S" time constants, if L_{AFmax} or L_{ASmax} are required. A potential source of error is aliasing, if the signal is not filtered to a bandwidth less than the sampling rate divided by the Nyquist number.

It is frequently stated that aliasing is avoided if the sample rate is 2 x the signal bandwidth, but that is only true if there is an anti-aliasing filter applied to the signal before sampling that has a "brick-wall" filter characteristic. In practice, no analogue filter is possible with an infinitely steep roll-off above its cut-off frequency, and allowance must be made for the rate of attenuation with frequency. The "span" of a filtered signal may be taken as 80% of the bandwidth, and the sampling rate should be a minimum of 2× the frequency of the bandwidth, i.e. 2/0.8 = 2.5. A value of 2.56 is commonly used in FFT analysis.

If the index which is to be used for assessment of the measured vibration is an r.m.s. index such as L_{AFmax} or L_{ASmax}, it is important to note that these indices involve exponential averaging of the root-mean-square value and are not the same as short-term energy averages. There can be a significant difference between, for example, L_{ASmax} (in which the "*S*" signifies the use of a 1-s time weighting) and $L_{Aeq,1s}$ or L_{AFmax} (in which the "*F*" signifies the use of a 125 ms time weighting) and $L_{Aeq,0.125s}$.

The effect of windowing of the signal prior to frequency transform is also important. It is common practice to use a Hanning window to deal with the problem of discontinuity at the ends of a window. This is because an FFT gives a spectrum for a signal that is an infinite chain of windows identical to the window being analysed, and if there is a step at the end where the start and finish of the window join, it will result in spurious components in the frequency spectrum. Application of a window to taper off the start and finish of the window avoids this effect, at the expense of frequency resolution. For a stationary signal, there will, however, be energy lost from the spectrum due to its attenuation at the start and finish, and this is allowed for by multiplying the frequency coefficients by $\sqrt{(8/3)}$ according to Parseval's theorem. However, if the signal is not stationary and it is of greater amplitude in the middle of the window than at the ends, this will overstate the amplitude by up to $10 \log_{10}(8/3) = 4.26$ dB. When an FFT spectrum is obtained using a "butterfly" algorithm it is important to remember that each line contains only half the amplitude for that frequency, and a correction of +6.02 dB is required. The RMS value of a fourier component is $1/\sqrt{2}$ (3.01 dB) less than the peak amplitude, so the overall correction to be applied to Hanning-windowed lines in a FFT spectrum is +7.27 dB.

UNCERTAINTY IN PREDICTION

Predictions of groundborne noise in cases where either the source does not yet exist in the proposed location, or the receiver has not yet been constructed, are subject to uncertainty for reasons that do not arise in the case of measurement uncertainty.

Source uncertainty

If the source is not yet in place, there is potential uncertainty of several kinds

- The source specification
- Installation specification
- Properties of the supporting/surrounding structures
- Properties of transmission media between the proposed source location and the receiver.

The source specification

The source of groundborne noise is usually a machine of some kind, and in many cases it is a railway. In all cases the mechanical properties of the source are clearly critical – in the case of machines these include not only its physical dimensions, masses and characteristics, such as number of blades or rotors, but also its operating parameters including speed, pressure, out-of-balance, roughness of rolling components and forces of impacts.

In the case of railways the important parameters are the wheel/rail roughness profile and other irregularities in the running surfaces, the masses of the suspension components from the unsprung bogie mass, the sprung mass of the bogie, the mass of the vehicle body and the dynamics of the vehicle body. The stiffnesses and damping characteristics of the suspension components are important, and may not be linear, and there may be stabilising links, and anti-roll bars attached with bearings having dynamic properties of their own. Of these, the key parameters are wheel/rail roughness and bogie unsprung mass that have a larger effect on source strength than any of the others.

Railway track characteristics are as important as those of the vehicles, including the dynamic stiffness of the rail pads and any resilient track support components such as resilient baseplates, booted blocks, booted sleepers or floating track slab.

At the time of prediction, some or many of these properties may be unknown, or not known in sufficient detail. In cases where similar rolling stock and/or track systems are installed elsewhere, measurements may be made to gather data from the existing cases, and it is then necessary to be able to make the required adjustments and corrections to allow for differences between the measured installation on the proposed installation. Those differences may include different tunnel shapes and sizes, different soils and intervening geology, and, of course, on top of uncertainty about the properties of source being measured and source being predicted, where measurement is involved, these are additional to measurement uncertainty.

In cases where the source already exists, measurement may be the first stage of the overall process, to be followed by prediction. It is rarely possible to measure at the vibration amplitudes and at exactly the location to be considered, and nearly always calculations have to be made to convert the measurement results to allow for structures to be built at the relevant location and the characteristics of the receptors.

Allowing for the effect of building a structure at the location concerned requires first, estimation of the magnitude of the coupling loss between the medium in which the vibration is measured and on or in which the structure will be built, and second, estimation of the response of relevant parts within the building where the vibration will be received by the sensitive receptor. If the quantity to be predicted is the sound level in the air within a room radiated from vibration surfaces in the room, the radiation efficiency of the surfaces and the influence of room acoustics, particularly room modes, significantly affect the answer. These topics are discussed in Chapters 6 and

8. A significant part of the uncertainty arising for these reasons is associated with selection of assessment location within the room. This may be laid down by specification, but if not, the result will have to be presented either as a maximum anywhere within the room, or as an energy average, or as a range.

Formal methods for assessing uncertainty

The Joint Committee for Guides in Metrology, set up by the leading standards authorities, publishes a "Guide to the expression of uncertainty in measurement" known as GUM.[1]

What the guidance document describes as a succinct summary of the procedure specified for evaluating and expressing uncertainty in measurement is as follows:

1. Express mathematically the relationship between the measurand Y and the input quantities X_i on which Y depends: $Y = f(X_1, X_2, \cdots, X_N)$. The function f should contain every quantity, including all corrections and correction factors, that can contribute a significant component of uncertainty to the result of the measurement.
2. Determine x_i, the estimated value of input quantity X_i, either on the basis of the statistical analysis of series of observations or by other means.
3. Evaluate the standard uncertainty $u(x_i)$ of each input estimate x_i. For an input estimate obtained from the statistical analysis of series of observations, the standard uncertainty is evaluated as described in the section "Type A evaluation of standard uncertainty". For an input estimate obtained by other means, the standard uncertainty $u(x_i)$ is evaluated as described in the section "Type B evaluation of standard uncertainty".
4. Evaluate the covariances associated with any input estimates that are correlated.
5. Calculate the result of the measurement, that is, the estimate y of the measurand Y, from the functional relationship f using for the input quantities X_i the estimates x_i obtained in step 2.
6. Determine the combined standard uncertainty $u_c(y)$ of the measurement result y from the standard uncertainties and covariances associated with the input estimates. If the measurement determines simultaneously more than one output quantity, calculate their covariances.
7. If it is necessary to give an expanded uncertainty U, whose purpose is to provide an interval $y - U$ to $y + U$ that may be expected to encompass a large fraction of the distribution of values that could reasonably be attributed to the measurand Y, multiply the combined standard uncertainty $u_c(y)$ by a coverage factor k, typically in the range from 2 to 3, to obtain $U = ku_c(y)$. Select k on the basis of the level of confidence required of the interval.

8. Report the result of the measurement y together with its combined standard uncertainty $u_c(y)$ or expanded uncertainty U. Describe how y and $u_c(y)$ or U were obtained.

GUM deals with uncertainty in measurement, and of course this has to be combined with uncertainty in prediction. A method that has been used is known as the Monte Carlo method, and is included in the Joint Committee for Guides in Metrology publication: JCGM 101:2008[2]

A Monte Carlo simulation involves running large numbers of simulations using randomly selected input parameters from a defined range and distribution. The process normally involves four steps:

1. Define the input range and probability distribution for modelling inputs
2. Create random input data sets based on the probability distribution of the inputs
3. Perform deterministic simulations of these sets
4. Collect and interpret the results, from which the error distribution of a prediction can be determined.

The Monte Carlo method has been applied in a number of cases.[3,4] Figure 12.1 is taken from Reference 4 and is typical of results obtained using the Monte Carlo method for groundborne noise prediction uncertainty. This

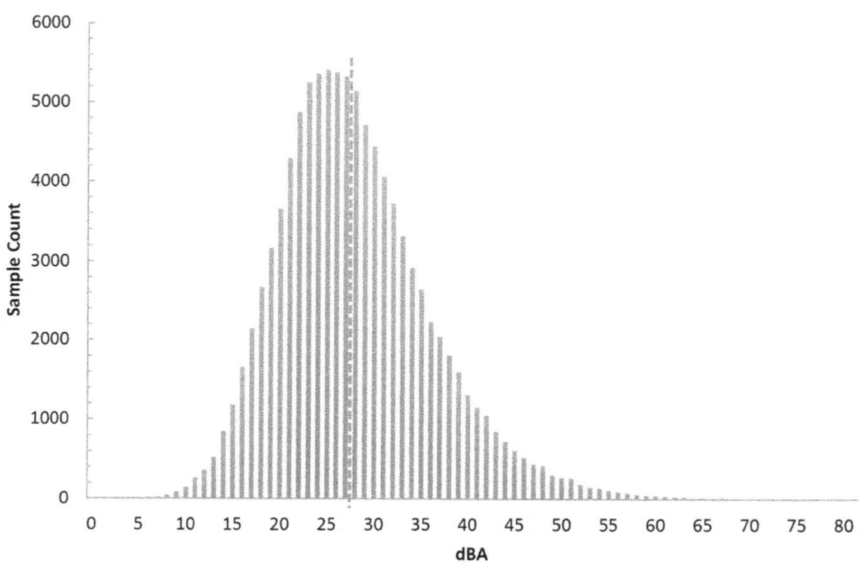

Figure 12.1 Distribution from 100,000 realisations of sample ground-borne noise prediction.

has a standard deviation of 8.2 dBA. The uncertainty calculated using the GUM uncertainty framework was 12.9 dBA, substantially greater than the uncertainty generated by the Monte Carlo simulation.

A wide range of software packages are available for implementing the Monte Carlo method.

NOTES

1. Guide 98-3. "Uncertainty of measurement—Part 3: Guide to the expression of uncertainty in measurement." (GUM:1995) ISO/IEC.
2. BIPM/JCGM 101. "Evaluation of measurement data—Supplement 1 to the Guide to the expression of uncertainty in measurement—Propagation of distributions using a Monte Carlo method." Joint Committee for Guides in Metrology, 2008.
3. Grau, L., and Laulagnet, B. "On the uncertainty of ground borne noise and vibration assessment." *Proceedings InterNoise 2019*, 2019.
4. Glickman, G., Faner, P., and Bell, K. "Applying the Monte Carlo method to rail transit ground-borne noise and vibration prediction." *Proceedings NoiseCon 2019 Institute of Noise Control Engineering*, 2019.

Chapter 13

Assessment criteria

As with all forms of noise, airborne and groundborne, a key stage in the assessment of groundborne noise is to determine the acceptability of noise levels. For airborne noise, there is a body of guidance which has led to well-established practices in assigning magnitudes of impact and degrees of significance to noise levels, although even with airborne noise there is lack of clarity when the noise is not from a transportation-related source, for example, when the noise is unwanted music from other premises or when the source is industrial in nature and possesses distinguishing characteristics.

With regard to groundborne noise, there is much less guidance, and what there is mainly relates to noise from underground railways. When the groundborne noise is caused by non-railway sources, such as construction work, mining and mineral extraction, machinery, gas flow in pipelines and electrical transmission installations, it is necessary to work from first principles.

When considering effects of groundborne noise, it is clearly necessary to take into account the nature of the receptor. In almost all cases it will be a human being, although the effect on animals and fish can arise in laboratories and in bodies of water such as lakes, rivers and aquaria. Noise in and around recording facilities may only indirectly involve effects on human beings. As far as effects on human beings are concerned, there are two main considerations. In circumstances where they may be sleeping, including the night period, effects of noise on sleep stages and awakening, and delay in falling asleep are the primary considerations. When sleep is not a concern, then the effects may be annoyance, effects on task performance or interference with speech communication.

For airborne noise, recognition has long been given to the fact that not only is the effect of noise on people dependent on frequency-weighted sound level, but also on duration and/or number of events in a given time period. Consequently, in addition to the set of noise scales which have been developed, principally the A-weighted decibel or dB(A), given the symbol L_A, a set of noise indices which are sensitive to time as well as level have become established. The first of these was the 10th percentile or the level

DOI: 10.1201/9781003618775-13

exceeded for ten per cent of the time L_{A10}. In the present day, the most widely used of these indices is the equivalent continuous sound level $L_{Aeq,T}$. The $L_{Aeq,T}$ index has become established for environmental airborne noise, mainly from transportation sources, following extensive noise and social surveys in which the correlation between measured noise and questionnaire answers was used to plot results such as the percentage of the population highly annoyed, against values of a composite noise index. The same approach has been used to study the effect of noise on sleep and on health indicators. Almost all these studies have been related to airborne noise and very few have been devoted to groundborne noise.

The A-weighting curve was originally created as part of a system intended to achieve weighted sound pressure level measurements that approximated the loudness perception of the human auditory system. Human hearing is much less sensitive to sound of low frequency and very high frequency, but that effect is level dependent, being greatest at low sound levels. Originally, three weighting curves were devised, "A", "B" and "C". The A curve was designed to follow the equal loudness curve at an A-weighted level of about 40 dB; the B curve was designed for B-weighted levels of around 70 dB and the C curve was for noise at a C-weighted level around 120 dB. Subsequently, it became widespread practice to use the A-weighting curve regardless of sound level, and as a result, when the sound level was significantly greater than 40 dB, the measurement applied excessive adverse weighting to low frequency sounds, and with regard to airborne sound levels, it is frequently criticised for that reason. However, when dealing with groundborne noise, particularly from railway sources, it is quite common for the sound levels concerned to be of the order of 40 dB, and at that level, as illustrated in Figure 13.1, the shape of the A-weighting curve is quite close to the shapes of other methods of allowing for the frequency response of the human ear. The first of these is contained in ISO 226[1] where the phon system of equal loudness curves is presented. The second is the Noise Rating (NR) system referred to below. It is apparent from this comparison that the shape of the A-weighting curve actually under weights sound in the 31.5 Hz one-third octave band by as much as 10 dB. A further feature of the low frequency weighting is that equal loudness curves are closer together at low frequencies than they are at higher frequencies, such that the often-stated dictum that "a 10 dB increase is approximately a doubling of loudness" is not correct at low frequencies. At 20 Hz, which is just into the audible range, a 5 dB increase changes the phon value by 10 units, whereas at 1000 Hz a 10 dB increase changes the phon value by 10 units.

It must be borne in mind that the phon scale in ISO 226 applies to pure continuous tones heard by otologically normal persons in the age range from 18 to 25 years and does not take into account the effect on loudness of a sound being either broad band or transient. There is another set of standards that provides methods of calculating the loudness in phons of a sound

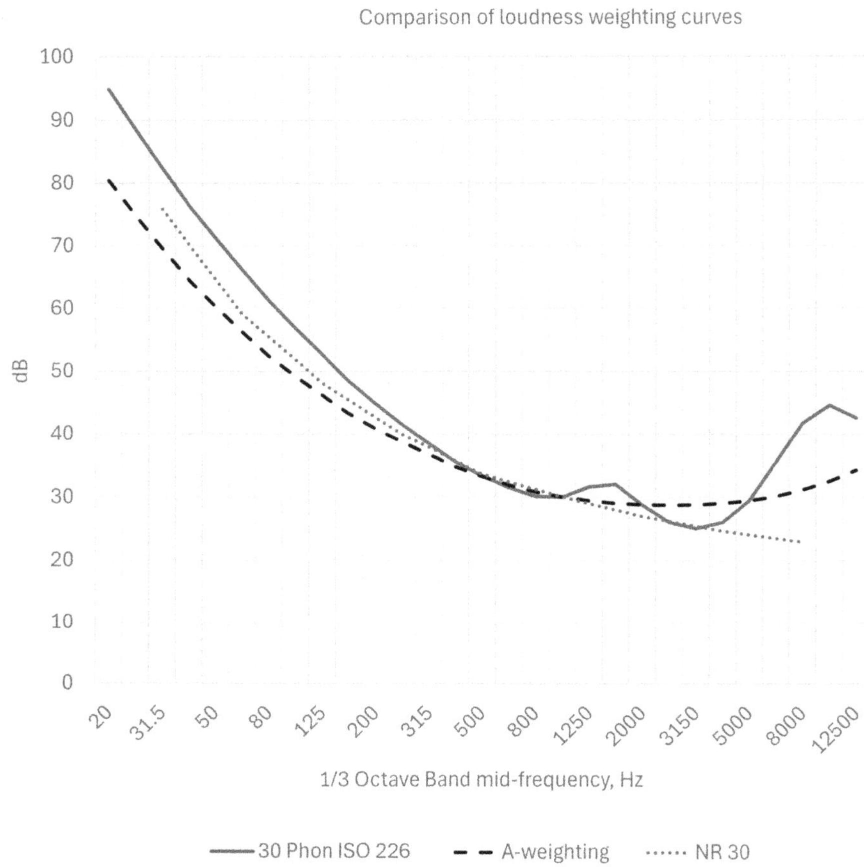

Figure 13.1 Comparison between the A-weighting curve, the 30 phon curve and the Noise Rating (NR) 30 curve.

spectrum evaluated in 1/3 octave bands, the ISO 532 series[2,3,4]. It should always be borne in mind that these standards present results for the mean of a distribution of hearing sensitivities which vary substantially among a population as can be seen from ISO 7029[5].

GROUNDBORNE NOISE FROM UNDERGROUND RAILWAYS

Groundborne noise from underground railways arises either when residents complain about existing railways or when a new railway is being planned and it is necessary to carry out an environmental impact assessment, and to persuade decision-makers that what is proposed is appropriate and acceptable.

When each new underground railway project is planned, the topic of groundborne noise design level criteria naturally comes under review. In the UK, the choice of criteria for groundborne noise dates from London

Underground's experience in responding to complaints about groundborne noise. While there have been railways under London for over 150 years, constructed in the main without mitigation against groundborne noise, the issue came to the fore when the Victoria Line was fully opened in 1969, and again when the Jubilee Line was opened between Baker Street and Charing Cross in 1979. These lines had the traditional track support of rails fastened to hardwood sleepers set into the tunnel invert, and there were complaints about groundborne noise from residents above. London Underground (LUL) carried out a set of measurements of maximum levels of A-weighted sound level inside dwellings. It has always been assumed that these results were obtained using the "slow" or "S" time weighting for exponential averaging of the RMS sound level, but that was not stated in the summary tables of measurement results. As the measurement results were manually recorded using a sound level meter with an indicating needle, reading a rapidly fluctuating needle position on "fast" response would have been more difficult than with "slow" response.

In recent times it has been the practice of LUL to use the L_{AFmax} scale rather than the L_{ASmax} scale when investigating complaints. A sound level meter or modern logging device set to "F" or "fast" response follows fluctuations in sound level much more readily than does a meter set to "S" or "slow" response. "F" response more closely matches the response of the human ear, while "S" response is more repeatable and predictable.

Where there are rail joints or discontinuities the difference between an "F" and "S" based measurement can be as much as 5 dB. Above an underground railway without rail joints and with resilient rail support the noise may be less peaky and the difference between "F" and "S" may be less.

In 1990 a report[6] by London Underground Ltd Professional Services group Scientific Services Division stated "From an analysis of the data it was concluded that complaints can be expected when the internal noise level exceeds 40 dB(A) max". This statement was relied on when the next new underground railway was planned in London, the Jubilee Line Extension.

A feature of most underground railways is that it is seldom necessary to consider intensification of an existing system. Assessment of development projects nearly always involves assessing new lines which give rise to groundborne noise in locations where there is no significant baseline – the few exceptions being where proposed new and existing lines cross or converge, but in those cases established practice is to assess groundborne noise from the new line in terms of absolute levels. In London, for example, in locations where there may be groundborne noise from several different underground railways, the approach is to design new underground railways without taking into account exceedance of L_{Amax} criteria by pre-existing railways, on the grounds that future improvements to their track support systems may result in reductions of the pre-existing noise levels.

The London Underground Noise and vibration asset design guidance, informed by the experience of LUL, states that for residential receptors, all reasonably practicable steps should be taken to design new subsurface railways so that the groundborne noise from the foreseeable level of service does not exceed 35 dB L_{AFmax}, the Lowest Observed Adverse Effect Level (LOAEL). The noise level representing the Significant Observed Adverse Effect Level (SOAEL) is stated as 40 dB L_{AFmax}.

NON-SOURCE-SPECIFIC GUIDANCE ON EFFECT OF MAXIMUM NOISE LEVELS

Design aims for groundborne noise can be considered in the context of national and international guidance on matters relating to noise and development. The Noise Policy Statement for England (NPSE), referred to by the National Planning Policy Framework advises

> There are two established concepts from toxicology that are currently being applied to noise impacts, for example, by the World Health Organisation. They are:
>
> NOEL – No Observed Effect Level
>
> This is the level below which no effect can be detected. In simple terms, below this level, there is no detectable effect on health and quality of life due to the noise.
>
> LOAEL – Lowest Observed Adverse Effect Level
>
> This is the level above which adverse effects on health and quality of life can be detected.
>
> 2.21 Extending these concepts for the purpose of this NPSE leads to the concept of a significant observed adverse effect level.
>
> SOAEL – Significant Observed Adverse Effect Level
>
> This is the level above which significant adverse effects on health and quality of life occur.
>
> 2.22 It is not possible to have a single objective noise-based measure that defines SOAEL that is applicable to all sources of noise in all situations. Consequently, the SOAEL is likely to be different for different noise sources, for different receptors and at different times. It is acknowledged that further research is required to increase our understanding of what may constitute a significant adverse impact on health and quality of life from noise. However, not having specific SOAEL values in the NPSE provides the necessary policy flexibility until further evidence and suitable guidance is available.

The first aim of the NPSE states that significant adverse effects on health and quality of life should be avoided, while also taking into account the guiding principles of sustainable development. The second aim is to mitigate and minimise adverse impacts on health and quality of life from environmental, neighbour and neighbourhood noise within the context of Government policy on sustainable development. The second aim of the NPSE refers to the situation where the impact lies somewhere between LOAEL and SOAEL. It requires that all reasonable steps should be taken to mitigate and minimise adverse effects on health and quality of life while also taking into account the guiding principles of sustainable development (paragraph 1.8). This does not mean that such adverse effects cannot occur.

Guidance on what noise level might be associated with the NOEL and LOAEL concepts in the case of airborne transportation noise is found in the Night Noise Guidelines for Europe[7] on page 99, in a discussion on health effects:

> However, noise-induced motility is a sign of arousal, and frequent (micro-) arousal and accompanying sleep fragmentation can affect mood and functioning next day and lead to a lower rating of the sleep quality. Therefore, motility is relevant for adverse health effects, but more than a few intervals with noise-induced motility are needed for inducing such effects. Although additional, more sophisticated analyses could be performed to refine this estimate, we propose $LAmax = 32$ dB(A) as the currently best estimate of the threshold for motility induced by transportation noise. The threshold found for EEG awakening was $LAmax = 35$ dB(A), that is, only a little higher than the 32 dB(A) found for noise-induced awakenings. This would mean that the *NOELAmax* for transportation noise events is most likely at most 32 dB(A), and definitely not higher than 35 dB(A). It is important to note that the above given $NOELAmax \sim 32$ dB(A) and $NOELAmax \sim 42$ dB(A) are indoor levels, in the sleeping room. Although events below 32 dB(A) are audible, and, hence, further research may show more sensitive effects than motility, on the basis of the present available evidence we propose to assume that $NOELAmax = 32$ dB(A) and set a health-based night-time noise limit that is tolerant for transportation noise events with *LAmax* 32 dB(A). On the other hand, since adverse health effects need to be prevented by health-based limits and even though vulnerable groups may require lower limits, on the basis of the present available evidence we propose to assume that $NOAELAmax = 42$ dB(A) and set a health-based night-time noise limit that does not tolerate transportation noise events with $LAmax > 42$ dB(A).

The Night Noise Guidelines go on to consider the concept of L_{night}, which is the annual energy average in terms of the L_{Aeq} index between the hours of 2300–0700.

> On the basis of the above proposal, it would be possible to derive a night-time noise guideline value in terms of L_{night}. Such a guideline value would indicate the level below which no short-term effects are to be expected that would lead to temporary reduced health or chronic disease. Such a guideline value needs to be compared with guideline values derived directly with a view to preventing temporary reduced health and chronic diseases. In particular, for self-reported sleep disturbance, which is an expression of reduced well-being and may be an indication of effects that could contribute to cardiovascular disease, exposure–effect relationships have been derived on the basis of an extensive set of original data from studies from various countries (Miedema, Passchier-Vermeer and Vos, 2003; Miedema, 2004). The percentage of people reporting high noise-induced sleep disturbance (%HS) levels off at 45 dB (A) but at a non-zero effect level. The remaining effect may be caused by events not incorporated in the exposure assessment and it appears that if all noise contributions would be incorporated in the exposure metric, high noise-induced sleep disturbance would vanish between 40 dB(A) and 45 dB(A), say at 42 dB(A). Since values found for other temporary reduced health effects or chronic diseases, in particular cardiovascular diseases, will be higher, and considering self-reported sleep disturbance as an adverse effect, this would suggest $L_{night} = 42$ dB(A) as the NOAEL to be compared with the value derived from the short-term effects. Note that this is an outdoor level, which would, assuming partly opened windows and an actual insulation of 15 dB(A), correspond to an indoor equivalent night-time sound level of 27 dB(A). The above discussion is based on motility, EEG awakenings, and conscious awakening. In addition, EEG micro-/minor arousals, and autonomic reactions have been discussed above.

The L_{Aeq} index, and therefore the L_{night} index, are functions of the noise level, duration and number of events.

In the UK, the High Speed Two (HS2) project sets impact and effect levels[8] of 35 dB L_{ASmax} as LOAEL and 45 dB L_{ASmax} as SOAEL for residential buildings.

Internationally, a range of limits and design specifications have been adopted for groundborne noise. Malmö Citytunnel was required by the Swedish Environmental Court not to exceed 30 dB L_{ASmax} (with 30 dB L_{AFmax} as a goal) in housing, hospitals and churches. Australian guidance is found in "Interim Guideline for the Assessment of Noise from Rail Infrastructure Projects" which provides for 40 dB $L_{ASmax(day)}$ and 35dB $L_{ASmax(night)}$. The Austrian Önorm S 9012 sets values for satisfactory protection of 40 dB L_{ASmax} for night, where the train service ceases for four of the night hours, and 35 L_{ASmax} otherwise. The figures for good protection are 5 dB lower. Italian regulations set a limit of 35 dB L_{ASmax}.. Norwegian Technical Regulation TEK 97 is satisfied if 32 dB L_{AFmax} is achieved in bedrooms. Västlänken in Gothenburg has a guideline value of 30 dB L_{ASmax} for bedrooms. Switzerland has a directive BEKS 1999 with a guideline value of 25 dB $L_{Aeq\ 1h}$ at night for new construction in residential areas. The US Federal Transit Administration has impact criteria of 35 dB L_{ASmax} for residences where there are more than 70 events per day, increasing by 3 dB where there are between 70 and 30 events per day and by 8 dB where there are fewer than 30 events per day. In Ireland, Metro Link and Dart Underground has a limit of 35 dB L_{ASmax} for residential areas.

A University of Gothenburg study[9] found that groundborne noise from railways in tunnels generally had a negative effect on self-reported sleep, with the majority of self-reported measures being worse following nights with 45 dB $L_{AF,max}$ trains than in quiet control nights.

With increasing noise level, a greater number of sleep parameters were negatively impacted, demonstrating higher levels of noise-induced sleep fragmentation and disruption. 35 dB $L_{AF,max}$ events caused a slight increase in heart rate, with no significant effects for any questionnaire or other biological measures. A noise exposure level of 40 dB $L_{AF,max}$ led to event-related elevations of heart rate (both ΔHRmax and HRA), a 13.2 min adjusted mean increase of intermediate N2 sleep and low levels of self-reported sleep disturbance by noise. During 45 dB $L_{AF,max}$ nights there were event-related elevations of heart rate (both heart rate and amplitude) and increased arousal probability, a 15.3 min adjusted mean reduction of deep sleep, a reduced number of sleep-stage changes and negative effects on the majority of self-reported sleep measures, including greater sleep disturbance by noise than in 40 dB nights. Furthermore, noise events with a higher frequency spectrum led to greater elevations of heart rate and increased arousal probability than the low frequency spectrum events.

Relative to the control, there were no differences in self-reported sleep in 35 dB $L_{AF,max}$ nights, and sleep disturbance by noise was only seen in 40 dB $L_{AF,max}$ nights. Following nights with 45 dB $L_{AF,max}$ noise, however, sleep was rated as worse than the control in 11 of the 16 measures. Self-reported sleep disturbance by groundborne railway noise may, therefore, manifest somewhere between 40 and 45 dB.

A Norwegian study[10] found that at $L_{pAFmax} = 32$ dB, 20% were slightly or more than slightly annoyed, and 4% were moderately or more than moderately annoyed.

CRITERIA FOR NON-RESIDENTIAL BUILDINGS

The British Council for Offices Guide to Specification contains the following section:

Ground-borne noise and vibration

> Sources of vibration external to the building can transmit to the building structure. Vibration in the building structure can be perceived by humans as either tactile vibration or, more commonly, as re-radiated noise.

Guidance on groundborne noise and vibration can be found in BS6472-1: 2008 and ANC Guidelines 'Measurement and Assessment of Groundborne Noise & Vibration'. In addition, it is suggested that train induced vibration should not result in re-radiated noise levels of more than the following:

Cellular Offices:	40 dB $L_{\mathrm{Amax(slow)}}$
Meeting Rooms:	40 dB $L_{\mathrm{Amax(slow)}}$
Open Plan Offices:	45 dB $L_{\mathrm{Amax(slow)}}$

These levels may be perceptible. Each case must be examined carefully accounting for the occupants' uses and expectations, background masking noise levels, extent and magnitude of levels, together with the number and duration of individual vibration events and – in the case of over-ground trains – any airborne noise intrusion from the same source.

Building vibration isolation can be costly and should have a cost/benefit/risk analysis. Audio simulations of the predicted groundborne noise levels, together with those of background masking noise, can be extremely worthwhile during the decision-making process, especially where they are near the threshold of acceptability.

Building vibration isolation is normally the only practicable mitigation measure and involves the introduction of resilient bearings between the substructure and superstructure. It is essential that special advice be sought for groundborne noise at an early stage to identify the structural, cost and programme implications.

The High Speed Two project in the UK sets impact and effect levels for non-residential buildings. The impact levels are 25 dB L_{ASmax} for large auditoria and concert halls; 30 dB L_{ASmax} for sound-recording and broadcasting studios, theatres and small auditoria; 35 dB L_{ASmax} for places of meeting for religious worship, courts, cinemas, lecture theatres, museums and small auditoria or halls; and 40 dB L_{ASmax} for offices, schools, colleges, hospitals, hotels and libraries.

The Sizewell C project in the UK included a proposal to use an existing branch line for construction trains for a period prior to the opening of a new dedicated line serving the site, and this involved the assessment of the combined effect of airborne and groundborne noise at dwellings close to the railway. The following criteria were adopted.

For nighttime (2300-0700), groundborne noise should be assessed using the L_{Amax} index by considering the decibel sum of both the groundborne noise and the airborne noise in the octave bands up to and including 125 Hz, using "S" rather than "F" time weighting, with windows fully closed, including the benefit of acoustic secondary glazing to the windows if present. Combined groundborne noise and airborne noise, using the approach set out in the preceding paragraph, should then be assessed against the LOAEL and SOAEL values for groundborne noise alone, namely 35 dB L_{ASmax} and 50 dB L_{Asmax} respectively.

For daytime (0700–2300), two cases should be considered: (A) where there is no pre-existing source of groundborne railway noise and (B) where there is existing groundborne railway noise.

In case (A) the same criteria established for nighttime should be applied for all times of day. In case (B) groundborne noise should be assessed using the $L_{Aeq\ 0700-2300}$ index by considering the decibel sum of both the groundborne noise and the airborne noise in the octave bands up to and including 125 Hz. The resulting value should then be assessed against an internal LOAEL of 25 dB $L_{Aeq\ 16h}$ and internal SOAEL of 40 dB $L_{Aeq\ 16h}$.

GROUNDBORNE NOISE FROM NON-RAILWAY SOURCES

A feature of noise from underground railways is that it is frequently heard as a "rumble", at least in the case of railways with a degree of resilience in their track support systems. The spectrum of the noise tends to be concentrated in the frequency range 30 Hz to 125 Hz for a long-established railway with rigid track support, and a more modern railway with a degree of resilient track support will cause groundborne noise concentrated in the range 30 Hz to 80 Hz or lower.

Groundborne noise from non-railway sources may differ not only in terms of its frequency content, but also in terms of what are called, in airborne noise assessment, "acoustic features". Acoustic features are defined, for airborne noise, in BS 4142[11] as tonality, impulsivity, intermittency and include characteristics that are "readily distinctive against the residual acoustic environment". BS 4142 is expressly applicable to the rating and assessment of outdoor airborne sound and has no stated applicability to groundborne noise. For sound to which it is applicable, corrections of up to 9 dB are made to the measured sound level, to allow for the additional annoyance value of such features (+3 to +9 dB for impulsivity; 0 to +6 dB for tonality; +3 dB for readily distinctive intermittency). ISO 1996-1,[12] which also has no express applicability to groundborne noise, recommends level adjustments to airborne noise of 5 dB for regular impulsive characteristics; 12 dB for highly impulsive and 3 to 6 dB for prominent tones. In the case of both these standards the corrections/adjustments are applied to noise measured using an L_{Aeq}-based index.

In addition to the L_A scale, there are alternative methods of quantifying the effect of noise, such as the Noise Rating (NR) system[13] which is a family of curves of levels in each octave band from 31.5 Hz to 8 kHz. The curves take account of the relative insensitivity of the human ear to low frequency, as does the A-weighting scale, but rather than summing the energy at each frequency after weighting, as is the case with the dB(A), the procedure with NR curves is that the highest curve reached by any of the octave band levels is the NR value of the spectrum in question. For many spectra, NR levels are approximately 5 dB less than the same spectrum measured in dB(A).

The method of determining NR values is set out in BS 8233.[14] However, BS 8233 states in Annex D.6 "Projects involving groundborne noise from underground trains, plant or industrial sources usually require expert advice."

Because non-railway groundborne noise is a less common phenomenon than airborne noise there are few precedents with regard to the application of limits. One of the causes of such groundborne noise, which has been the subject of limitation by planning condition, is the excavation of rock from underground mines. Another non-railway source of groundborne noise is the operation of a tunnel boring machine (TBM).

Common sources of structureborne noise in buildings are lifts, and DIN Standard 8989 includes limiting values of acceleration in the lift shaft wall from the point of view of controlling structureborne noise levels from lifts.

Significance criteria for underwater noise

Fish

Underwater sound is expressed using a different decibel scale to that used for airborne sound, with a reference level of 1μPa rather the 20μPa in the case of airborne sound. Furthermore, the characteristic impedance of water is much greater than that of air, resulting in underwater sound levels being numerically much higher than would be the case if the water were replaced by air.

Fish can sense waterborne noise through both auditory and mechanosensory systems.[15] Hearing threshold curves for various species of fish have been obtained and over the mid-frequency range are typically about 75 dB re 1μPa or higher. Most research has centred on the relationship between sound pressure levels and hearing damage in fish. Sounds that were lower than 180 dB (re 1 μPa) and sounds that were not on continuously had no apparent impact on the sensory cells of the ear.[16]

It is not known what effect noise has in the range between 75 and 180 dB, on such matters as the behaviour of fish in, for example, swimming to their spawning grounds. Rivers and oceans are, however, naturally noisy, with heavy rain producing sound pressure levels up to 110 dB. Shipping can produce noise levels over 105 dB.

Fish species also have pressure-sensitive cells that are sensitive to low frequencies (typically 10 Hz to 30 Hz) and near-field pressure changes, limited to an area immediately surrounding the fish. This allows the fish to detect the presence of other fish in close proximity (such as in schooling behaviour) or assists in predatory avoidance. The sensitivity of the pressure-sensitive cells, however, is at an insufficient level to detect tunnel boring machine operation.

Fish are known to have complex and diverse inner ear structures with a broad range of hearing sensitivities, generally weighted towards lower frequencies. Some species, such as some members of the herring family, can hear across a wider range of frequencies and quieter sounds; these species are sometimes referred to as hearing specialists. Others, particularly flatfish, such as dab and plaice, hear over a narrower range of frequencies and have much lower sensitivities to sound (they have no swim bladders); salmon and trout fall into this latter group also, which are sometimes referred to as hearing generalists. Some examples of fish hearing thresholds[17] are given in Figure 13.2.

Popper et al.[18] groups species of fish according to whether they possess a swim bladder, and whether it is involved in its hearing. This guidance gives specific criteria, as both SPLpeak (the peak sound pressure) and SELcum values, for a variety of noise sources. The modelled criteria are summarised in Table 13.1. SELcum is defined as the sound exposure level over a number of individual impulsive sound exposures and is calculated as the log sum of

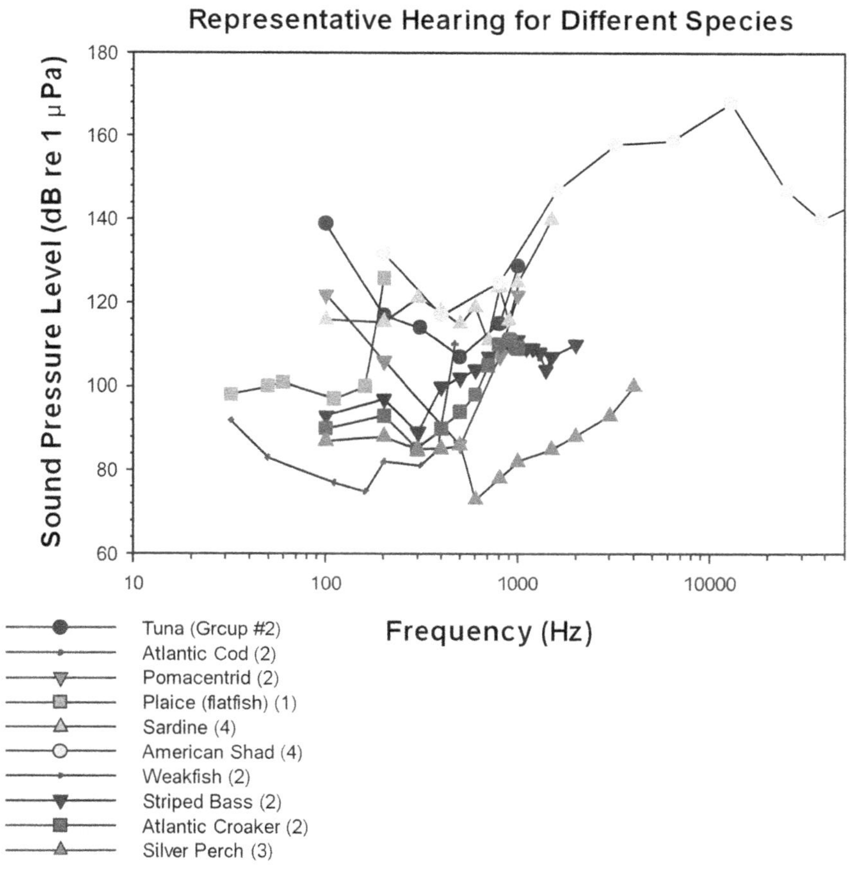

Figure 13.2 Examples of fish hearing thresholds.

Table 13.1 Criteria for assessment of mortality and potential mortal injury, recoverable injury and temporary threshold shift (TTS) in species of fish (Popper et al., 2014)

Fish: no swim bladder	>219 dB SELcum or >213 dB SPLpeak	>216 dB SELcum or >213 dB SPLpeak	>>186 dB SELcum
Fish: swim bladder is not involved in hearing	210 dB SELcum or >207 dB SPLpeak	203 dB SELcum or >207 dB SPLpeak	>186 dB SELcum
Fish: swim bladder involved in hearing	207 dB SELcum or >207 dB SPLpeak	203 dB SELcum or >207 dB SPLpeak	186 dB SELcum

the squared sound pressure of the individual events. It is primarily applicable to impulsive sound from, for example, percussive piling.

Popper et al. also consider behavioural effects in fish, which are defined as "substantial change in behaviour for the animals exposed to a sound. This may include long-term changes in behaviour and distribution, such as moving from preferred sites for feeding and reproduction, or alteration of migration patterns."

The Popper et al. (2014) guidelines conclude that there is insufficient data available to apply quantitative thresholds for behavioural effects on fish.

Mammals

It is important to bear in mind the fact that the A-weighting curve only applies to human beings. In cases where animals are affected by noise, their significantly different hearing thresholds must be taken into account, as A-weighted sound levels (dB(A)) are not relevant to animals. There is no equivalent to the "A-weighting" for those cases; instead, the frequency content of the sound has to be considered as a spectrum rather than as an overall single number.

For mice, hearing thresholds are broadly similar to those of humans but with the frequency scale multiplied by 10 to give maximum sensitivity between 10 kHz and 40 kHz (compared with 1 kHz and 4 kHz for human beings). The effect of this is that, while humans hear sounds such as whistles as being the loudest for any given physical magnitude, and hear rumbles and very high pitched hisses at the same physical magnitude as being much quieter, mice are most sensitive to very high pitched sounds right at the top of the hearing range of humans with undamaged hearing, and extending up into the ultrasound region, inaudible to human beings. A young human being with undamaged hearing can just hear a sound at 20 kHz, but at half the loudness (or less) than a whistle at 1 kHz, and the A-weighting reduces the measured level of a 20 kHz sound by 9.2 dB. This is the upper limit of the frequency range of most sound level meters, although some have ultrasound bands centred on 31.5 kHz and 40 kHz, and this is the region where mouse hearing is highly sensitive.

HORSES

Horses have significantly different hearing thresholds such that their perception of noise is different from that of humans. At frequencies which are predominant in noise from transportation sources the hearing of horses is much less sensitive than that of humans. Horses can sense very low frequency noise and vibration by the transmission of vibration from the ground to the middle ear via the hooves, teeth and jawbone.

The interpretation of the significance of noise levels on humans can only be done through the means of combined social and noise surveys, which cannot be done for animal species. Reaction to noise at environmental levels, by any species depends as much on the meaning of the noise as on its level. For example a factor in determining human annoyance to noise is emotional response on the basis that the maker of the noise is doing something without proper regard for others and/or for their own private benefit to the detriment of others. Humans experience annoyance due to task interference. Animals do not experience such reactions. Task interference could arise in equitational activities, where vocal aids are given to the horse by the rider, but the level of noise would have to be at a high enough level to cause speech interference after allowing for the different hearing threshold of the horse. Animals are most sensitive to sounds made by other animals particularly of their own species (and, in the case of predator species, the sounds made by their prey), and to sounds which are associated with predators themselves. Sounds that convey meaning including, for example, the imminent arrival of feed, also attract attention.

Horses may be startled by what they see as well as what they hear. The type of noise event which can potentially startle a horse is one involving a sudden onset, and a short rise-time for the sound level. Modern horses are very used to noise of relatively high level, particularly those involved in competitive events, as the locations of these may have high sound levels from public address systems, from other activities at the same event and from transportation sources, as well as crowds. There are helipads at most major racecourses. In most yards at equine establishments large machinery and lorries come and go throughout the day, and in arable areas combine harvesters and other agricultural equipment come close to horses with negligible effect. It is common to see horses grazing with a train travelling at speed close by their field.

Marine mammals

The National Marine Fisheries Service guidance (NMFS, 2016) for marine mammals covers underwater noise and its effects. The guidance groups marine mammals into functional hearing groups and applies filters to the unweighted noise to approximate the hearing sensitivity of the receptor. The hearing groups given in the NMFS (2016) are summarised in Table 13.2 and Figure 13.3. A further group for Otariid Pinnipeds is also given in the

Table 13.2 Marine mammal hearing groups

Hearing group	*Example species*	*Generalised hearing range*
Low Frequency (LF) Cetaceans	Baleen Whales	7 Hz to 35 kHz
Mid-Frequency (MF) Cetaceans	Dolphins, Toothed Whales, Beaked Whales, Bottlenose Whales (including Bottlenose Dolphin)	150 Hz to 160 kHz
High Frequency (HF) Cetaceans	True Porpoises (including Harbour Porpoise)	275 Hz to 160 kHz
Phocid Pinnipeds (PW) (underwater)	True Seals (including Harbour Seal)	50 Hz to 86 kHz

Source: From NMFS, 2016.

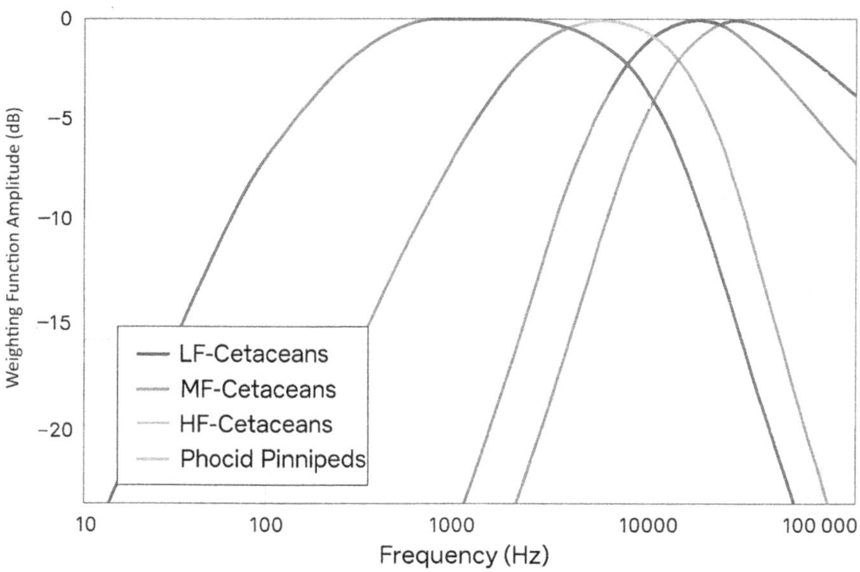

Figure 13.3 Auditory weighting functions for low frequency (LF) Cetaceans, mid-frequency (MF) Cetaceans, high frequency (HF) Cetaceans, and Phocid Pinnipeds (PW) (underwater).

(From NMFS, 2016.)

guidance for sea lions and fur seals, but this has not been used in this study as those species of pinnipeds are not found in this region.

NMFS (2016) presents single strike, unweighted peak criteria (SPLpeak) and cumulative (i.e., more than a single sound impulse), weighted sound exposure criteria (SELcum) for both permanent threshold shift (PTS), where unrecoverable hearing damage may occur, and temporary threshold shift (TTS), where a temporary reduction in hearing sensitivity may occur in individual receptors. Table 13.3 presents the NMFS (2016) criteria for onset of risk of PTS and TTS for each of the key marine mammal hearing groups.

Where SELcum are required, a fleeing animal model has been used. This assumes that the animal exposed to high noise levels will swim away from the noise source. For this a constant fleeing speed of 1.5 ms^{-1} has been assumed, which is a cruising speed for a harbour porpoise. These are considered "worst case" as marine mammals are expected to be able to swim much faster under stress conditions. The model assumes that a fleeing receptor stops if it reaches the coast before the noise exposure ends.

Criteria for disturbance or behavioural reaction effects in marine mammals are in development by NMFS. For this assessment, thresholds as single strike SEL have been derived from data for mid-frequency and for high frequency cetaceans. The disturbance threshold for seals is as per TTS. Criteria have not been presented for low frequency cetaceans (Table 13.4).

Criteria for sensitive equipment

Guidance on acceptable levels of vibration for different forms of sensitive equipment and operation were originally published by Bolt Beranek and Newman (BBN)[19] and Ungar, Sturz and Amick.[20] The BBN curves have

Table 13.3 Criteria for assessment of auditory injury and TTS in marine mammals (NMFS, 2016)

	Unweighted SPL_{peak} (dB re 1 μPa)	*Weighted SEL_{cum} (dB re 1 $\mu Pa^2 s$)*	
NMFS (2016)	*Auditory injury (PTS)*	*Auditory Injury (PTS)*	*TTS (Temporary Threshold Shift)*
Low Frequency (LF) Cetaceans	219	183	168
Mid-Frequency (MF) Cetaceans	230	185	170
High Frequency (HF) Cetaceans	202	155	140
Phocid Pinnipeds (PW) (underwater)	218	185	170

Table 13.4 Criteria for assessment of disturbance/behavioural reaction in marine mammals

Hearing group	*Behavioural reaction SEL re 1 $\mu Pa^2 s$*
Mid-Frequency (MF) Cetaceans	160 dB
High Frequency (HF) Cetaceans	145 dB

been widely used but a need to apply more stringent criteria for some spaces has since emerged. The BBN criteria were repeated in the ASHRAE Guidelines[21] and the latest version of the criteria, known as vibration criterion (VC) curves, can be found in the IEST Recommended Practice 012.3[22] and this version of the criteria is reproduced in Table 13.5 and Figure 13.4.

Two criteria, VC-F and VC-G have been added to the original set of curves. These are used for the evaluation of existing facilities with extremely vibration sensitive conditions.

In addition to the VC system is the guidance in NIST-A of the National Institute of Standards and Technology (NIST) in Gaithersburg, Maryland, which is more stringent than VC-E at frequencies below 20 Hz.

Table 13.5 Application of vibration criterion curves

Criterion curve	*Description of use*
Workshop (ISO)	Distinctly perceptible vibration. Appropriate to workshops and non-sensitive areas.
Office (ISO)	Perceptible vibration. Appropriate to offices and non-sensitive areas
Residential day (ISO)	Barely perceptible vibration. Appropriate for sleep areas. Usually adequate for computer equipment, probe test equipment and microscopes less than 40X.
Operating theatre (ISO)	Vibration not perceptible. Suitable in most instances for microscopes to 100X and for other equipment of low sensitivity.
VC-A	Adequate in most instances for optical microscopes to 400X, microbalances, optical balances, proximity and projection aligners, etc.
VC-B	Appropriate for inspection and lithography equipment (including steppers) to 3 μm line-widths.
VC-C	Appropriate standard for optical microscopes to 1000X, lithography and inspection equipment (including moderately sensitive electron microscopes) to 1A good standard for most lithography and inspection equipment to 1 μm detail size.
VC-D	Suitable in most instances for demanding equipment including many electron microscopes (SEMs and TEMs) and E-Beam systems.
VC-E	A challenging criterion to achieve. Assumed to be adequate for the most demanding of sensitive systems including long path, laser-based, small target systems, electron-beam lithography systems working at nanometer scales, and other systems requiring extraordinary dynamic stability.
VC-F	Appropriate for extremely quiet research spaces; generally difficult to achieve in most instances, especially cleanrooms. Not recommended for use as a design criterion, only for characterisation.
VC-G	Appropriate for extremely quiet research spaces; generally difficult to achieve in most instances, especially cleanrooms. Not recommended for use as a design criterion, only for characterisation.

As measured in one-third octave bands of frequency over the frequency over the frequency range 8–80 Hz (VC-A and VC-B) or 1 to 80 Hz (VC-C through VC-G).

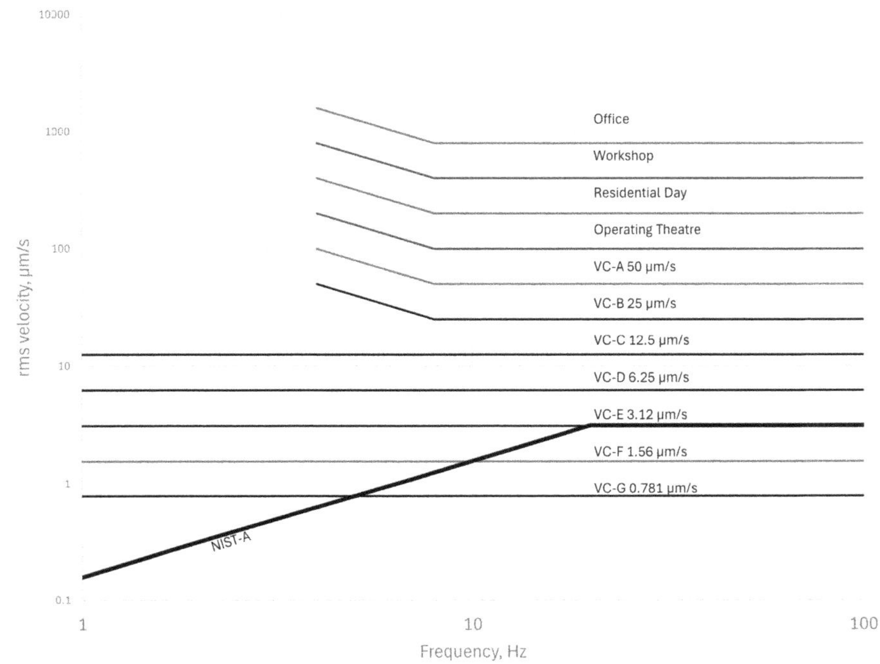

Figure 13.4 Graphical definition of vibration criterion and NIST-A curves.

The NIST-A criterion is identical to VC-E at frequencies above 20 Hz but sets a constant rms displacement amplitude at lower frequencies. It corresponds to a displacement of 0.025 μm between 1 and 20 Hz and a velocity of 3.1 μm/s between 20 Hz and 100 Hz (note that VC curves end at 80 Hz).

The level of 0.75μm/s indicated for NIST-A1 is very close to the level of 0.78 μm/s represented by VC-G.

Both the VC and NIST criteria are assessed based on rms vibration velocity in one-third octave bands, but there is no clear guidance as to how the one-third octave band magnitudes are assessed and analysed for a transient signal.

For vibration limits on sensitive equipment, manufacturer's information may be available. Limit details vary between different types of equipment and operation and may be given as a single figure or may define a particular frequency range.

NOTES

1. ISO. "Acoustics—Normal-equal-loudness-level contours." ISO 226:2003, International Organization for Standardization, Geneva, 2003.
2. ISO 532-1:2007 – Acoustics – Methods for calculating loudness – Part 1: Zwicker Method, International Organization for Standardization, Geneva, 2007.

3. ISO 532-2:2017 – Acoustics – Methods for calculating loudness – Part 2: Moore-Glasberg Method, International Organization for Standardization, Geneva, 2017.
4. ISO 532-3:2023 – Acoustics – Methods for calculating loudness – Part 1: Moore-Glasberg-Schlittenlacher Method, International Organization for Standardization, Geneva, 2023.
5. ISO 7029:2017+A1:2024 – Acoustics – Statistical distribution of hearing thresholds relating to age and gender, International Organization for Standardization, Geneva, 2024.
6. Bovey, E.C. "Statement to noise standards working party." London Underground Ltd Professional Services group scientific services division, 6 June 1990.
7. World Health Organization. *Night Noise Guidelines for Europe*. WHO Regional Office for Europe, Copenhagen, 2009.
8. High Speed Two (HS2) Limited. "Information paper E10: Control of ground noise and vibration from the operation of temporary and permanent railways." High Speed Two (HS2) Limited, Version 2 6 July 2022.
9. Smith, M. G., Ögren, M., Morsing, J.A., and Persson Waye, K., "The effects on sleep of ground borne noise from trains in tunnels." *Proceedings Euronoise 2018*, 2018.
10. Aasvang, G.M., Engdahl, B., and Rothschild, K., "Annoyance and self-reported sleep disturbances due to structurally radiated noise from railway tunnels." *Applied Acoustics*, 68(9), 970–981, September 2007.
11. BS4142:2014+A1:2019. *Methods for Rating and Assessing Industrial and Commercial Sound*, British Standards Institution, London, 2019.
12. ISO 1996-1 2016. *Acoustics—Description, Measurement and Assessment of Environmental Noise —Part 1: Basic Quantities and Assessment Procedures*, 3rd ed., International Organization for Standardization Geneva, 3 January 2016.
13. Kosten and van Os, "Community reaction criteria for external noises." *National Physical Laboratory Symposium*, No. 12, 1962, 377, London H.M.S.O.
14. BS 8233:2014. *Guidance on Sound Insulation and Noise Reduction for Buildings*, British Standards Institution, London, 2014.
15. Clark, J.A., Young, J.A., Bart, A., and Zohar, Y., "Physiological effects of infrasonic noise on captive fish." *Journal of the Acoustical Society of America*, 100(4), 2709, Woodbury, NY, October 1996.
16. Hastings, M.C., Popper, A.N., Finneran, J.J., and Lanford, P.J., "Effects of low-frequency underwater sound on hair cells of the inner ear and lateral line of the teleost fish Astronotus ocellatus." *Journal of the Acoustical Society of America*, 99(3), 1759–1766, March 1996.
17. Nedwell, J.R., Edwards, B., Turnpenny, A.W.H., and Gordon, J. *Fish and Marine Mammal Audiograms: A Summary of Available Information*. Prepared by Subacoustech Ltd., Hamphire, UK. Report 534 R 0214, 2004.
18. Popper, A.N., Hawkins, A.D, Fay, R.R., Mann, D.A., Bartol, S., Carlson, T.J., Coombs, S., Ellison, W.T., Gentry, R.L., Halvorsen, M.B., Løkkeborg, S., Rogers, P.H., Southall, B.L., Zeddies, D.G., and Tavolga, W.N. *Sound Exposure Guidelines for Fishes and Sea Turtles*. Springer Briefs in Oceanography, 2014. 10.1007/978-3-319-06659-2
19. Gordon, C., Saunders, M., and Cooper, D. *Vibration: A Quantitative Approach to Building Design*. Stanford University, Stanford, California, November 23, 1988.

20. Ungar, E.E., Sturz, D.H., and Amick, C.H. "Vibration control design of high technology facilities." *Sound and Vibration*, 24(7), 20–27, 1990.
21. ASHRAE. *Sound and Vibration Control*, The American Society of Heating, Refrigerating and Air-Conditioning Engineers (ASHRAE): HVAC Applications (SI Edition), 2007.
22. Institute of Environmental Sciences and Technology IEST-RP-CC012.4 Contamination Control Division Recommended Practice 012.3, Considerations in Cleanroom Design, 2015.

Glossary

Accelerometer: A transducer which generates a charge proportional to vibration acceleration, usually used in connection with a charge amplifier.

Acoustic: In the context of groundborne noise, audible or with significant spectral content within the audible range of 20Hz to 20 kHz.

Alluvium: Soil transported by water in suspension and deposited by sedimentation.

Anisotropic: Having different properties in different directions.

A-weighting: A filter applied to the measured or calculated spectrum of an audible signal to approximate the unequal frequency response of human hearing.

Bending stiffness: For a beam, the product of Young's modulus E and second moment of area I of a section. For a plate see **Flexural rigidity**.

Blow count: Used in connection with the Standard Penetration Test (SPT). It is the number of blows to penetrate 305mm (12 inches) of soil. See **Standard Penetration Test.**

Boulder clay (also till or glacial till): A mixture of clay, sand, gravel and boulders remaining after an ice sheet has melted.

Bulk modulus: An elastic constant relating mean stress to volumetric strain.

Charge amplifier: An amplifier which converts a charge signal from an accelerometer into a voltage signal, with sensitivity in millivolts per coulomb.

Clay: A fine-grained geological material with particle sizes less than 2 microns and practically impervious.

Coefficient of permeability: A measure of the rate of flow through soils under a unit hydraulic gradient and laminar flow at standard temperature and pressure.

Coherence: When two or more signals have a fixed phase relationship.

Coherent line source: A one-dimensional continuous row of coherent sources so that amplitude has the same phase along the whole length of the source.

Complex number: A number with a real and imaginary part, thus, describing the amplitude and phase angle of a time-varying magnitude.

Compression: The decrease in volume of a medium.

Consolidation: The time-dependent process of reducing the volume of a soil (soil settlement) by expulsion of water.

Coulomb damping: A constant damping force caused by friction between moving surfaces, which is independent of frequency and velocity.

Crest factor: The ratio of the peak of a time-varying signal to its RMS value.

Decibel (dB): A scale that denotes the ratio between two quantities that is proportional to power, the number of decibels being ten times the common logarithm of this ratio, or between the squares of two amplitudes such as pressure or velocity in which case it is 20 times the common logarithm of the amplitude ratio.

Density (mass): The mass in kilogrammes per cubic metre of a medium.

Dirac function (also known as a delta function): An impulse that is zero at all times except at a single infinitely short point in time, where it is infinitely large and for which $\int_{-\infty}^{\infty} \partial(x)\partial t = 1$. Its integral over time = 1. Its frequency spectrum has an amplitude of 1 at all frequencies (see also **Kronecker delta**).

Dispersion: A property of wave motion in which phase speed increases with frequency.

Elasticity: The ability of a body to regain its original geometry after imposed loads or stresses are removed.

Euler's formula: $e^{ix} = \cos(x) + i\sin(x)$.

Fast: see **Root-mean-square (RMS)**.

Fast Fourier transform (FFT): A discretised form of the Fourier transform coded so as to facilitate numerical solution in a short time.

Flexural rigidity: $Eh^3/(12(1-\sigma^2))$ where E is Young's modulus; h is plate thickness and σ is Poisson's ratio.

Fourier transform: Decomposition of a time-varying signal into a spectrum of sine waves.

Friction damping: see **Coulomb damping**.

Geophone: A transducer that generates a voltage signal proportional to vibration velocity.

Gravel: Particles of rock greater than 4.75 mm.

Green's or Green Function: The impulse response of a system being the transfer function representing the response of a receiver to an input in the form of a Dirac function.

Ground beam: A concrete beam connecting pile caps in the foundations of a building.

Group velocity: The propagation speed of the envelope of a wave packet which results from frequency dependence of wavenumber. $V_g = \mathrm{d}\omega/\mathrm{d}k$, where ω is angular frequency and k is wavenumber.

Hann window (also Hanning window): A time weighting applied to a window of a digitised signal which tends to zero at the beginning and end in order to remove spurious frequency components caused by the fact that a Fourier transform assumes that the data within the analysis window is repeated indefinitely.

Head wave (also a von Schmidt wave): A wave which has a depth-dependent horizontal propagation speed, being equal to the p-wave speed at the ground surface, falling to that of the shear wave speed at a depth below ground, where its wavefront forms a tangent to the shear wavefront.

Hysteretic damping: Damping resulting from a lag between stress and strain in a vibrating medium or system.

Imaginary number: A number multiplied by $\sqrt{-1}$.

Impedance: The complex ratio of force to velocity, or force per unit area (pressure) to velocity.

Incoherence: When two or more signals have a random phase relationship.

Isotropic: Having the same properties in all directions.

Kronecker delta: A discrete version of a Dirac delta function.

Lamb wave: An elastic wave that propagates in plates.

Line source: A one-dimensional continuous row of incoherent sources so that amplitude has random phase along the whole length of the source.

Line source transfer mobility: The ratio of velocity at a receiving point to force per unit length at a line source.

Lithology: The characteristics of rock formation, often extended to include other geological formations.

Loam: A mixture of sand, silt and clay that may contain organic material.

Loess: A windblown, uniform fine-grained soil.

Logarithmic decrement: The natural logarithm of the ratio of the amplitudes of successive peaks in the decay curve that results from an impulse excitation, or the instantaneous removal of strain, as in a step function.

Loss factor: The ratio of the imaginary to the real part of a modulus, when expressed as a complex number.

Love wave: A surface wave with horizontal motion perpendicular to the direction of wave propagation.

MDOF: Multiple degree of freedom.

Mobility: The inverse of **Impedance.**

Modulus of elasticity: The ratio of stress to strain for an elastic material.

Moment of inertia: see **Second moment of area.**

Mud: Clays and silts mixed with water into a viscous fluid.

Nyquist frequency: The highest frequency that can be accurately represented in a digitally-sampled signal, equal to half or 1/2.56 times the sampling frequency.

Octave and one-third octave: A series of frequency bands used to describe the level of sound or vibration across a frequency spectrum. An octave encompasses a range in which the upper frequency limit is double the lower frequency limit. A one-third octave represents a range in which the upper limit is $2^{1/3}$ times the lower limit.

Oedometer: A device to apply vertical loads to a soil.

Orthotropic: Having properties that are different in three mutually perpendicular directions.

Overburden: The soil pressure at a depth of intertest.

Overconsolidated soil: Soil that has experienced past stresses greater than its current stress.

Overconsolidation ratio: The ratio of the past effective stress (preconsolidation stress) to the current effective stress.

Peak level: The maximum instantaneous sound or vibration magnitude (positive or negative) over a defined time period.

Peak particle velocity (PPV): The maximum (positive or negative) instantaneous velocity at a point on a vibrating object. Usually expressed in mm/s.

Permeability: The rate of flow of water through soils.

Phase speed: The propagation speed of a wavefront of constant phase.

Pile: A long, slender structural element used to transfer structural loads deep within the soil.

Pile cap: A concrete block on top of a pile, usually supporting a floor slab.

Plasticity: The property of a material in which the material does not recover its original geometry after imposed stresses are removed. It is also the consistency of a soil in which moulding of the soil does not produce cracks.

Plasticity index: The range of water content within which soil can be moulded without cracking.

Poisson's ratio: The negative ratio of transverse strain perpendicular to an applied force to axial strain, or strain in the direction of the force.

Pore water pressure: The pressure of the water within voids in a soil.

Porosity: The ratio of the volume of voids to the total volume of soil.

Power spectral density: Power per unit frequency or wavelength interval.

Power sum: Square root of the sum of the squares of a series or sequence of values.

Preconsolidation pressure: The effective stress that a soil was subjected to in the past.

p-wave: Compression wave

Quality factor: The inverse of the loss factor.

Rayleigh wave: A wave on the surface of an elastic half-space (at the interface between an elastic solid and a void) with an elliptical motion and phase speed approximately 90% of the shear wave velocity.

Re-radiated noise: Noise arising within internal spaces as a result of the vibration of surfaces bounding space. The vibration source may be inside a building (structureborne noise), or separated from the building by the ground (groundborne noise).

Rock: A rigid mineral mass.

Root-mean-square (RMS): The square root of the mean of the squares of the instantaneous sound or vibration over a specified time period. For a time-varying signal exponential averaging is employed using either a $\tau = 0.125$ second ("Fast" or "F") or $\tau = 1$ second ("Slow" or "S") time constant the equation is $L_\tau = \sqrt{\frac{1}{\tau} \int_{\xi=0}^{t} e^{\frac{t-\xi}{\tau}} x^2(\xi) d\xi}$ where x is the time-varying signal.

Sand: Hard soil particles between 4mm and 0.075mm.
Saturation: A condition in which the voids in a soil mass are filled with water.
Scholte wave: Similar to a Rayleigh wave at the interface between an elastic solid and a fluid rather than a void.
SDOF: Single degree of freedom.
Secant modulus: The slope of a stress strain graph drawn between its origin and a specific point on the graph.
Second moment of area: The product of an area and the square of the distance of its centroid from a specified axis. Used in the calculation of bending stiffness, given the symbol *I*.
Shear stress: The force per unit area that acts parallel to a plane in a material causing it to deform by sliding or shearing.
Silt: A fine soil particle with sizes within the range 0.075 mm and 0.002 mm.
Slow: see **Root-mean-square.**
Slow wave: A second type of compression wave predicted by the equations of Biot and Stoll as existing in fluid-filled, poro-elastic media.
Soil: A geological material derived from the weathering of rocks.
Standard Penetration Test (SPT): A test which gives the resistance of a soil to the energy delivered on a standard sampler by a hammer of weight 140 lbs falling from a height of 30 inches.
Stiffness: The ratio of force to displacement of a spring-like element.
Stoneley wave: An interface wave that, under limited conditions, propagates along the boundary between two elastic media.
Strain: A change in a material's dimensions caused by an applied stress.
Stress: Force per unit area.
Superposition: The combination of two or more signals by arithmetic addition of their time-varying amplitudes.
s-wave: Shear wave
Tangent modulus: The slope of the stress-strain curve at a specific point representing the material's instantaneous stiffness, that is, the first derivative of the stress-strain relationship.
Till: see **Boulder clay.**
Transfer mobility: The ratio of velocity at a receiving point to force at its source.
Vibration velocity level: (VdB) 20 $\log(v/v_{ref})$. Two conventions regarding v_{ref} are in use, 1 nm/s and 50 nm/s.
Viscosity: The measure of a fluid's resistance to flow due to internal friction.
Viscous damping: A force that opposes motion and is proportional to velocity and is frequency dependent.
Void: The space between soil particles.
von Schmidt wave: see **Head wave.**
Wave number: $k = 2\varpi/\lambda = (\omega/c)$ where λ is wavelength and c is phase speed.

WAV file: A digital audio file that stores one or more uncompressed digital signals in either 16-bit, 24-bit or 32-bit format, including header information.
Weathering: A process in which rocks disintegrate into soils.
Young's modulus: The ratio of tensile stress to tensile strain.

LIST OF SYMBOLS

i, j	$\sqrt{-1}$
b, C	viscous damping, Ns/m
c	phase velocity, wave speed, m/s
c_g	group velocity, m/s
c_r	Rayleigh wave speed, m/s
c_s	shear wave speed, m/s
η, β	loss factor dimensionless
F	force, N
G	shear modulus, N/m^2
k	wave number $2\varpi/\lambda$, 1/m
l	length, m
m	mass, kg
m', ρ'	mass per unit length, kg/m
m'', ρ''	mass per unit area, kg/m^2
ρ	mass density, kg/m^3
S	cross-sectional area, m^2
Z	impedance, Ns/m

Index

Pages in *italics* refer to figures, pages in **bold** refer to tables, and pages followed by "n" refer to notes.

For Product Safety Concerns and Information please contact our EU representative GPSR@taylorandfrancis.com
Taylor & Francis Verlag GmbH, Kaufingerstraße 24, 80331 München, Germany

www.ingramcontent.com/pod-product-compliance
Ingram Content Group UK Ltd.
Pitfield, Milton Keynes, MK11 3LW, UK
UKHW020201200926
475702UK00004B/557
9781041019978